Lifetime Reliability-aware Design of Integrated Circuits

Mohsen Raji • Behnam Ghavami

Lifetime Reliability-aware Design of Integrated Circuits

 Springer

Mohsen Raji
Shiraz University
Shiraz, Iran

Behnam Ghavami
Shahid Bahonar University of Kerman
Kerman, Iran

ISBN 978-3-031-15347-1   ISBN 978-3-031-15345-7   (eBook)
https://doi.org/10.1007/978-3-031-15345-7

This Springer imprint is published by the registered company Springer Nature Switzerland AG
The registered company address is: Gewerbestrasse 11, 6330 Cham, Switzerland

Preface

In recent decades, integrated circuits have achieved great performance and power improvements owing to aggressive manufacturing technology downscaling. However, smaller feature sizes and higher power densities (and consequently temperature) have made modern VLSI circuits suffer from day-to-day increasing wear and aging effects. Hence, the lifetime reliability of integrated circuits has become a serious and growing challenge for the industry. So, it is necessary to analyze and improve the lifetime reliability of digital circuits such as flip-flops and combinational circuits as the basic components of large-scale integrated circuits.

This book investigates the design techniques for modern electronic systems which are mainly used in applications requiring high reliability and safety such as space satellites, automotive, or healthcare electronic systems. In this book, automated design approaches are presented to analyze and improve the lifetime reliability of electronic systems. The emphasis is on modeling techniques and computer-aided design (CAD) algorithms for the analysis and improvement of lifetime reliability issues in nano-scale CMOS integrated circuits.

Shiraz, Iran   Mohsen Raji
Kerman, Iran   Behnam Ghavami

Acknowledgment

After a decade of researching and teaching in the field of fault-tolerant computing systems, we decided to write this book. The groups of students from Shiraz University and Shahid Bahonar University of Kerman who developed the mathematical analysis approaches, design algorithms, and software packages have made the most important contributions to this work.

The authors would like to thank Professor Hossein Pedram for keeping us engaged and learning. Our special thanks go to our graduate students Atousa Jafari, S. Milad Ebrahimipour, and Reza Mahmoudi who developed the materials for chapters and helped with the preparation of this book. All this would not have been possible without their valuable support.

Writing this book was harder than we thought and more rewarding than we could have ever imagined. None of this would have been possible without our families.

June 2022
Mohsen Raji
Behnam Ghavami

Contents

Impacts of Process Variations and Aging on Lifetime Reliability of Flip-Flops

1 Introduction

Despite various advantages in energy and performance parameters, aggressive technology scaling has led to new design challenges to the reliability of complementary metal-oxide-semiconductor (CMOS) digital circuits [1]. Nanoscale circuit has encountered different sources of variabilities such as process variations (PV) [2] and aging-induced variability [3] which significantly result in discrepancies between the circuit operation during design, after manufacturing, and after its operation. The limitations of CMOS chip manufacturing tools and the physical constraints in nanometer dimensions have led to process variabilities, including random dopant fluctuation (RDF) [4] and line edge roughness (LER) [5]. On the other hand, aging of circuit components, in particular negative- and positive-bias temperature-induced instability (NBTI and PBTI) respectively occurring in p-channel metal-oxide-semiconductor (PMOS) and n-channel metal-oxide-semiconductor (NMOS) transistors, have been considered as the most important lifetime reliability challenges of technologies of less than 45 nm [6]. NBTI and PBTI mainly increase the threshold voltage of transistors over time, thereby increasing the delay of circuit transistors and gates, which, in turn, may lead to violation of circuit delay constraints [6].

Flip-flops (FFs) are known to be the key building components of digital circuits and systems as they have a considerable contribution on the performance-energy tradeoff at the chip level [7, 8]. Typically, FFs have undeniable effects on the chip's robustness, considering the continuous increase in larger process variabilities [8]. For example, aging and PV-induced variations that violate maximum delay constraints result in performance overhead and timing yield loss due to increased clock cycle period [9]. Furthermore, with emerging and increasing of transistor aging impacts, the lifetime reliability of FFs became a challenge for modern very-large-scale integration (VLSI) systems [10]. So a deep understanding of the impact of variations and aging on FFs is required to enable a selection of FF topology for modern, highly reliable chip designs.

© The Author(s), under exclusive license to Springer Nature Switzerland AG 2023
M. Raji, B. Ghavami, *Lifetime Reliability-aware Design of Integrated Circuits*,
https://doi.org/10.1007/978-3-031-15345-7_1

Until now, various research studies have been performed to comparatively analyze the different characteristics of FFs, considering the challenges at the nanoscale level. However, each analysis has limitations in the challenges taken into account. Some of these works consider the impact of PV in the designing of FFs, while others only focus on NBTI or BTI without considering PV effects. There are also a few research works that have considered aging mechanisms when analyzing the characteristics of FFs, but they are not specifically dedicated to analyzing the lifetime reliability of FFs.

There are a few works that have taken account of both PV and aging when comparing different FFs. In Ref. [11], a comparative analysis of the effects of the different sources of variations (PV as well as BTI) on the performance of FFs is carried out. In Ref. [12], the resilience and yield of two representative FF in CMOS technology are investigated, taking account of the impact of process variations and aging using transistor-level Monte Carlo (MC) simulations. In Ref. [13], the effects of PV and NBTI/PBTI aging on the write noise margins of various metal-oxide-semiconductor field-effect transistor (MOSFET) and fin field-effect transistor (FinFET)-based FF cells are comparatively analyzed. However, the timing reliability of FFs is neglected in that study. Despite the valuable results, the mentioned works mainly suffer from a lack of comprehensiveness: they either neglect PBTI effects or PV or focus on other parameters such as static or write noise margins, ignoring timing issues concerning FFs that are needed to be addressed for lifetime reliability. More importantly, there is not a well-proven and formal metric to fairly evaluate and compare the timing reliability of FFs while taking account of the impacts of process variations and aging mechanisms.

In this chapter, a comprehensive comparative analysis of the timing reliability of the different types of FFs is carried out, taking into consideration the impact of process variations and aging mechanisms [14]. To this end, a novel metric called timing-yield-aware lifetime reliability (TYR) is proposed, in which the timing yield of an FF determined by the user/designer is considered as the reference point for the analysis of its lifetime reliability. Based on TYR, two different classes of FF topologies are comparatively analyzed: master-slave and pulsed FFs. In the analysis, the effect of process variations on transistor channel length and width is taken into account. Moreover, both NBTI and PBTI mechanisms are considered by shifting the transistor voltage threshold of FFs. Based on TYR and using extensive Monte-Carlo-based simulations, it is shown that the PV and aging mechanisms significantly decrease FF timing reliability by 40% in some cases. It is also shown that for a 95% timing yield and with a variation ratio of 15% as well as 3 years of operation time, the lifetime reliability of pulsed FFs decreases to 63% while master-slave FFs' lifetime reliability degrades to 56%. Moreover, the obtained results indicate that the lifetime reliability of FFs decrease with an increase in the amount of PV and an increase in operational lifetime. There are some FFs, like charge control FFs (XCFFs), that are more reliable than others under process and runtime variations. However, some FFs, such as write port master-slave (WPMS) flip-flops and semi-dynamic flip-flops (SDFFs) are more vulnerable to aging effects. The proposed metric and methodology can be used as a well-defined guideline for FF topology selection in making high-reliable circuit designs.

The rest of the chapter is organized as follows: Section 2 presents the method of analysis used in this study, in which the FF topologies under study are introduced and then flip-flop timing characteristics and the process variation and aging models are presented. In Section 3, the proposed TYR metric is described. In Section 4, the lifetime reliability of FFs is analyzed based on TYR, and the impacts of PV and/or aging mechanisms on the different FFs under various conditions are investigated. Finally, Section 5 concludes the chapter.

2   Analysis Methodology

In this section, the methodology used for analysis is described. The circuit topology of the two main categories are presented and then, the timing parameters of FFs which have the main roles in the correct functionality of FFs are defined. Moreover, the modeling approach of PV as well as the aging effects are explained.

2.1   Flip-Flop Topologies Under Study

The circuit topologies of FF under study are divided into two main classes: master-slave FFs and pulsed FFs. The schematic circuit designs of the FFs are shown in Fig. 1. The selected FFs are as follows:

1. Master-Slave FFs [15]:

   - Transmission gate FF (TGFF) (Fig. 1a)
   - Modified transmission gate FF (TGFFv2) (Fig. 1b)
   - Transmission gate master-slave FF (TGMSFF) (Fig. 1c)
   - Write port master-slave FF (WPMS) (Fig. 1d)

2. Pulsed FFs [16]:

   - Hybrid-latch FF (HLFF) (Fig. 1e)
   - Cross-charge control FF (XCFF) (Fig. 1f)
   - Semi-dynamic FF (SDFF) (Fig. 1g)
   - UltraSPARC semi-dynamic FF (USDFF) (Fig. 1h)

2.2   Timing Parameters of Flip-Flops

There are timing parameters that play a prominent role in the correct functionality of FFs [17]. For a correct functionality of a given FF, the data should be stable for a

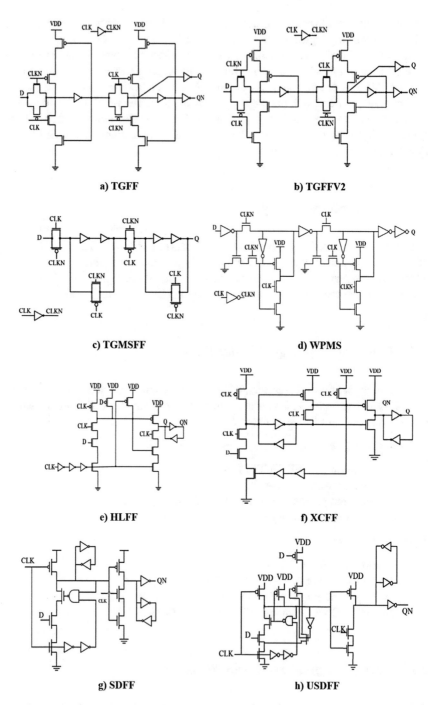

Fig. 1 Topologies of FFs under study

Fig. 2 Timing parameters
of an FF

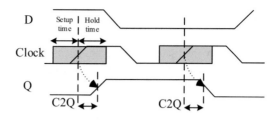

Fig. 3 Sequential circuit
timing path

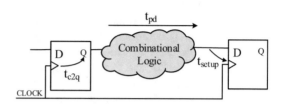

specific time duration (called as *setup time*) before the sampling edge of the clock which is. Also, the data should remain stable and unchanged for a specific time duration (called *hold time*) after the sampling edge of the clock. Clock-to-output delay (C2Q) is the delay from the sampling clock edge until the data are latched at the output. Depending on the logical value of the latched data at the FF output, we define 0-to-1 C2Q (1-to-0 C2Q) as the delay from the sampling clock edge to a low-to-high transition (high to low transition) at the FF output. On the other hand, data-to-output delay (D2Q) is the delay from data arrival at the input until the data are latched at the FF output Q. Typically, D2Q is defined as the sum of setup time and C2Q [10, 17]. Figure 2 shows the different timing parameters for FFs.

Both digital sequential circuits and pipeline circuit stages consist of combinational logic and FFs. As a result, FFs have an important effect on the determination of circuit design parameters, such as power consummation, delay, and reliability. Delay of a sequential circuit or a pipeline stage is the superposition of the time which data passes from a launching FFs and then, passing through the combinational logic and after that, being captured by a state-storing FF. Figure 3 demonstrates a representative schematic diagram of the sequential/pipeline stage circuit timing path. In order to work correctly, the following inequality should be established among existing timing quantities:

$$t_{setup} + t_{pd} + t_{c2q} < t_{clk}$$

where t_{setup} is the FF setup time, t_{pd} is the combinational circuit propagation delay, t_{c2q} shows the C2Q delay of the FF, and t_{clk} is the clock period. As a result, t_{c2q} and t_{setup} have significant roles in the correct functionality of the circuit, and their variations may lead to a reduction in the reliability of FFs and the sequential circuit consequently. Therefore, without the loss of generality, we consider the parameter D2Q (which is the summation of C2Q and setup time [10, 17]) in investigating the lifetime reliability of the FFs under study.

2.3 Aging Effects

Many individual physical mechanisms may contribute to aging effects in MOSFET transistors, such as NBTI, PBTI, hot carrier instability (HCI), and time-dependent dielectric breakdown (TDDB). They affect the reliability of circuits by increasing the absolute value of the transistor threshold voltage, resulting in increased delays in gates and paths. As the most important aging mechanisms, NBTI and PBTI occurring in PMOS and NMOS transistors respectively, briefly referred as BTI, are wear-out mechanisms which results in transistor timing degradation when the transistor is in static state. BTI contributes to increasing the threshold voltage (V_{th}) when the transistor is in a stress condition; i.e., the gate voltage (V_g) of PMOS and NMOS transistors are $V_g =$ "0" and $V_g =$ "1," respectively, and the gate-to-source voltage is $|V_{gs}| = V_{dd}$. Stress condition is straightly related to the signal probability of the gate of the transistor, i.e., the relative duration in which $V_g =$ "0" and $V_g =$ "1" in PMOS and NMOS, respectively.

2.4 BTI Model

Increased V_{th} due to BTI (ΔV_{th_BTI}) is obtained using the following equation [11]:

$$\Delta V_{th\_BTI} = a * (TSP.t)^n \tag{1}$$

where a is a technology- and temperature-dependent constant (obtained from long-term prediction models based on R-D method for specific CMOS technology and environmental conditions), t is the time, n is the BTI time exponential constant, and TSP is the transistor stress probability (the ratio between the time duration in which the transistor is under BTI stress to the total time). TSP is a function of the signal probability of inputs that reflects the stress condition of a transistor in a circuit during total operation time.

2.5 Process Variation Model

Aggressive technology scaling has led to large PVs originating from random dopant fluctuations (RDFs), line-edge roughness (LER), and oxide thickness fluctuations. Experimental and simulation-based results have indicated that the standard deviation of V_{th} (σ_{vth}) caused by PV is related to transistor channel width and length [11]:

$$\sigma_{vth} = \frac{A_0}{\sqrt{WL}} \tag{2}$$

where A_0 is a technology-dependent parameter and W and L are, respectively, the channel width and length of the transistor. Equation 2 implies that the variation of V_{th} is universely proportional to the square root of the transistor active area.

3 V_{th} Degradation Analysis Approach

Threshold voltage degradation is analyzed considering both PV and aging effects. First, the aging-induced threshold voltage degradation of each transistor is calculated using Eq. 1, taking into consideration its temperature and temperature-sensitive parameter (TSP). The parameters of the BTI model are configured such that the worst case of BTI-induced delay degradation becomes 10% over 3 years for a simple inverter [11, 18]. In order to consider PV-induced V_{th} degradation in Monte Carlo simulation, a set of ΔV_{th} is generated for all FF transistors based on their channel widths and lengths in accordance with Eq. 2. MC simulation parameters are set such that V_{th} variation (i.e., $3\sigma/\mu$, which is the ratio between the standard deviation and the mean of V_{th}) for a minimum-sized transistor at 16-nm technology nodes becomes equal to 10%, 15%, and 20% [19]. V_{th} degradation for each transistor ($\Delta V_{th}(\text{total})$) is computed as:

$$\Delta V_{th}(\text{total}) = \Delta V_{th}(\text{aging}) + \Delta V_{th}(PV) \tag{3}$$

where $\Delta V_{th}(PV)$ shows the threshold voltage degradation caused by PV and $\Delta V_{th}(\text{aging})$ is the aging-induced threshold voltage degradation. After calculating ($\Delta V_{th}(\text{total})$) for each transistor in a FF, the combined analysis of the impact of variability sources are measured.

4   Timing Yield-Aware Lifetime Reliability Metric

In order to have a formal metric for the fair comparison of the lifetime reliability of FFs, in this section, we describe the proposed metric—the timing yield-aware lifetime reliability (TYR) metric. In order to derive TYR metric, we begin with the definition of lifetime reliability of a FF as shown in Eq. 4. The lifetime reliability of an FF at time τ ($R(\tau)$) is defined as the probability of an FF to be operational at time τ (i.e., at the end of its lifetime), given that it works correctly at time 0 (i.e., in fresh time); that is:

$$\mathcal{R}(\tau) = P(\text{operational at } \tau | \text{ operational at } 0) \tag{4}$$

Since the D2Q parameter of an FF is a common metric to analyze its quality of operation from the timing point of view [20], this parameter is selected as a FF's delay in computing TYR of a FF. So in order to have an operational FF at time 0, the value of the D2Q parameter should be less than the timing constraint ($\mathcal{T}$). To be reliable at the end of lifetime, the D2Q of the FF at time τ should be less than the same timing constraint at fresh time ($\mathcal{T}$). Hence, Eq. 4 can be rewritten as follows:

$$\mathcal{R}(\tau) = P(D_\tau < \mathcal{T} \mid D_0 < \mathcal{T}) \tag{5}$$

where D_τ and D_0 respectively represent the values of the D2Q parameters for an FF at time τ and time 0 and $\mathcal{T}$ is the timing constraint of the FF, which can be either defined by the designer at the design stage or computed considering other timing constraints of the design, such as the critical path delay of the combinational part.

In order to solve Eq. 5 and find lifetime reliability, we use the conditional probability calculation rules [21]. Hence, we have:

$$\mathcal{R}(\tau) = \frac{P(D_\tau < \mathcal{T} \cap D_0 < \mathcal{T})}{P(D_0 < \mathcal{T})} \tag{6}$$

where $P(D_\tau < \mathcal{T} \cap D_0 < \mathcal{T})$ shows the probability of the intersection of two events, $D_\tau < \mathcal{T}$ and $D_0 < \mathcal{T}$ (i.e., the value of D2Q parameter for FF in time τ and time 0 be less than the timing constraint $\mathcal{T}$), and $P(D_0 < \mathcal{T})$ indicates the probability of $D_0 < \mathcal{T}$ occurring.

Since the delay parameters (including D2Q) increase by increasing the lifetime of FFs [20, 22], we have:

$$D_0 < D_\tau \tag{7}$$

Therefore, if event $D_\tau < \mathcal{T}$ occurs, event $D_0 < \mathcal{T}$ has certainly occurred. So we have:

$$D_0 < \mathcal{T} \subset D_\tau < \mathcal{T} \tag{8}$$

where $\subset$ shows that event $D_0 < \mathcal{T}$ is a subset of event $D_\tau < \mathcal{T}$. Based on the rules of probability calculation, we have [21]:

$$P(D_\tau < \mathcal{T} \cap D_0 < \mathcal{T}) = P(D_\tau < \mathcal{T}) \tag{9}$$

Considering Eq. 9 in Eq. 6, lifetime reliability is obtained as follows:

$$\mathcal{R}(\tau) = \frac{P(D_\tau < \mathcal{T})}{P(D_0 < \mathcal{T})} \tag{10}$$

Due to the impacts of variabilities, the timing parameters of FFs, including the D2Q parameter, are modeled as random variables with Gaussian distribution [22]. So $P(D_0 < \mathcal{T})$ is equal to the value of the cumulative distribution function (CDF) of D2Q at time 0 for $\mathcal{T}$. In its turn, $P(D_0 < \mathcal{T})$ is equal to the timing yield of

FF, which is set by the designer at the design stage. Therefore, we can reexpress Eq. 10 as:

$$\mathcal{R}_{\mathcal{Y}}(\tau) = \frac{\phi_\tau\left(\phi_0^{-1}(\mathcal{Y})\right)}{\mathcal{Y}} \tag{11}$$

where $\mathcal{Y}$ shows the timing yield of FF (equal to $P(D_0 < \mathcal{T})$), $\phi_0^{-1}(\cdot)$ indicates the inverse cumulative distribution function (ICDF) of D2Q at time 0, $\phi_\tau(\cdot)$ shows the CDF of D2Q at time τ, and $\mathcal{R}_{\mathcal{Y}}(\tau)$ is the lifetime reliability of FF at time τ, taking into account timing yield $\mathcal{Y}$ determined at the design stage. Please note that we changed the notation of lifetime reliability from $\mathcal{R}(\tau)$ to $\mathcal{R}_{\mathcal{Y}}(\tau)$ to distinguish between different timing yield levels, which can be chosen by the designer or computed considering other timing constraints of his/her design. Equation 11 shows the formula for computing the TYR metric for an FF with timing yield $\mathcal{Y}$.

The TYR degradation for a specific $\mathcal{Y}$ ($\Delta\mathcal{R}_{\mathcal{Y}}(\tau)$), which means the fraction of FFs which have been operational in time 0 but failed in time τ to the total number of FF operational in time 0 is computed as:

$$\Delta\mathcal{R}_{\mathcal{Y}}(\tau) = \frac{\mathcal{Y} - \phi_\tau\left(\phi_0^{-1}(\mathcal{Y})\right)}{\mathcal{Y}} * 100 \tag{12}$$

The procedure for calculating the TYR for FFs is shown in Fig. 4. The data include the circuit netlist, PV/aging information, and the timing yield goal ($\mathcal{Y}$) for the FF design set by the designer. Monte Carlo HSPICE simulations are performed for PV-aware and then, the distribution parameters (μ and σ) of FF D2Q for the fresh time are extracted from simulation. The ICDF of D2Q ($\phi_0^{-1}(\mathcal{Y})$) is computed to determine the timing constraint ($\mathcal{T}$) which should be satisfied by the design to achieve the timing yield goal. Meanwhile, PV- and aging-aware Monte Carlo HSPICE simulations are performed to extract the distribution parameters (μ and σ) of FF D2Q for the operation time. In this step, considering timing constraint T, D2Q CDF ($\phi_\tau\left(\phi_0^{-1}(\mathcal{Y})\right)$) is computed, and finally, the lifetime reliability of the FF ($\mathcal{R}_{\mathcal{Y}}(t)$) and lifetime reliability degradation ($\Delta\mathcal{R}_{\mathcal{Y}}(t)$) are calculated using Eqs. 11 and 12, respectively.

5   Experimental Results

In this section, we conduct some experiments to investigate the impacts of process variations and aging on the lifetime reliability of FFs under study using the proposed TYR metric. So, in the following, the characterization setup is explained and then, FF characterization results in design time stage (i.e. time 0) is provided. Then the FFs are comparatively analyzed based on the TYR metric, including the lifetime reliability analysis of FFs. Moreover, the impacts of aging/process variations on the reliability of FFs are investigated. Also, the power-delay-product of FF topologies are reported for different FF topologies.

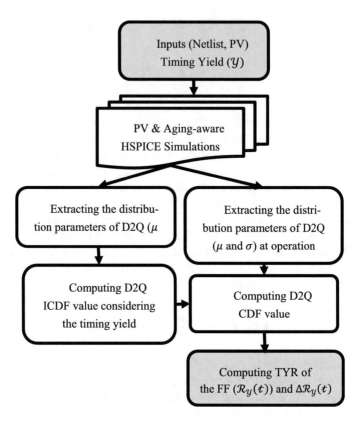

Fig. 4 The procedure for calculating the TYR for FFs

5.1 Characterization Setup

Figure 5 shows the simulation setup circuit for each FF under study. Two-stage buffer with minimum sized inverters are used for data and clock inputs. A capacitor with capacity 2fF is also used as the load in the output. This loading allows realistic running conditions to evaluate performance parameters. The FFs are implemented and simulated in HSPICE using predictive technology model (PTM) library 16 nm LP [23]. The temperature and supply voltage are 25 °C and a 0.9 volts for nominal conditions, respectively.

5.2 FF Characterization Results

Table 1 shows the basic characteristics of eight different FFs (from both master-slave and pulsed categories) at design time (i.e., without considering the impacts of process variations and aging). The first two columns of Table 1 respectively show

Fig. 5 The simulation setup

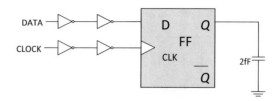

Table 1 Characterized data for the different FFs under study

FFs	# of transistor	Transition	C2Q$_{min}$(ps)	Setup time (ps)	Static power (nw)	Dynamic power (fW/MH)	Average power (µw)
TGFF	22	0 to 1	116.2	41	7.58	0.63	0.057
		1 to 0	133.4	27		1.57	
TGFFv2	22	0 to 1	122.5	45	6.27	0.88	0.062
		1 to 0	142.4	20		1.61	
MSFF	18	0 to 1	86.7	37	0.17	0.48	0.036
		1 to 0	51.3	42		1.21	
WPMSFF	26	0 to 1	102.4	40	9.06	1.80	0.16
		1 to 0	67.4	140		4.41	
HLFF	20	0 to 1	40.9	−3	4.73	3.12	0.28
		1 to 0	39.7	−5		5.81	
XCFF	21	0 to 1	75.1	−5	2.70	2.36	0.23
		1 to 0	112.2	−6		5.01	
SDFF	25	0 to 1	66.3	−1	4.11	1.08	0.73
		1 to 0	27.9	−2		1.68	
USDFF	27	0 to 1	70.6	−7	4.23	2.18	0.50
		1 to 0	25.2	−15		1.14	

the name and transistor count of the FF. The third column indicates the transition type of FF output (either 0 to 1 or 1 to 0). Some important characterization parameters of FFs, including C2Q$_{min}$, setup time, static and dynamic power, and average power are reported in the rest of the columns. Static power is measured when the inputs and outputs of circuits are stable, and it is obtained from the product of the leakage current and source voltage for each transistor in the circuits. Dynamic power is calculated when the circuit is switching. In order to measure C2Q$_{min}$, the value of the input data should be sufficiently settled before the clock edge. Setup time is measured by closing up the data to the active clock edge until the point where C2Q delay is increased by 10% with respect to the nominal C2Q delay (i.e., $1.1 \times$ C2Q$_{min}$). The negative values of the setup time in pulsed FFs show that the data input can be settled after the active clock edge. D2Qmin is calculated by summing up C2Q$_{min}$ and the obtained setup time. In order to analyze the impacts of transistor aging on FF delay, a MOSFET model reliability analysis (MOSRA) [24] tool is used in the experiments. Process variation impacts are investigated using HSPICE Monte Carlo simulations with 1000 runs. In these experiments, the process variations on transistor channel length and width are taken into consideration which are modeled by a normally distributed random variation. The constants in Eqs. 1 and

Table 2 Comparison of delay parameters for different FFs under process variation and aging at V_{dd} = 0.9

D2Q$_{min}$(ps)	Transition	FFs	Delay of process variation only (ps)		Delay of aging only (ps)	Delay of both process variation and aging (ps)	
			μ	σ		μ	σ
TGFF	0 to 1	157.2	157.1	6.8	162.4	162.9	24.3
	1 to 0	160.4	162.0	9.7	168.8	169.5	35.3
TGFFv2	0 to 1	167.5	167.1	7.4	171.3	171.6	25.6
	1 to 0	162.4	163.1	11.1	169.8	170.4	36.1
TGMSFF	0 to 1	123.7	124.5	5.2	129.1	129.4	17.4
	1 to 0	93.3	92.5	3.1	94.9	95.3	10.1
WPMSFF	0 to 1	142.4	143.1	7.9	148.2	148.6	37.3
	1 to 0	207.4	201.8	7.1	212.7	213.2	50.2
HLFF	0 to 1	37.9	37.4	2.6	39.7	40.1	8.1
	1 to 0	34.7	35.6	1.7	35.8	36.1	6.8
XCFF	0 to 1	70.1	70.1	5.7	70.7	71.1	6.3
	1 to 0	106.2	106.9	8.3	108.2	108.6	16.2
SDFF	0 to 1	65.3	65.3	6.1	68.3	68.8	21.5
	1 to 0	25.9	27.5	2.1	28.7	29.1	7.2
USDFF	0 to 1	63.6	63.8	4.2	65.5	66.2	9.3
	1 to 0	10.2	11.5	4.5	15.9	16.1	6.2

2 assume the following values: $n = 1/6$, $a = 3.9*10^{-3}$ [16], and $A_0 = 10^{-9}$ [12], respectively.

Table 2 includes information about the delay parameter of each FF in different states, such as design time, variation only with a variation ratio ($3\sigma/\mu$) of 15%, aging only (lifetime is considered 3 years), and both aging and process variation. As can be seen, process variation and aging have an effect on delay parameters. These results are calculated at a supply voltage of 0.9v.

5.3   Aging Impacts on Lifetime Reliability

In this subsection, we take advantage of TYR to analyze the impacts of aging on the lifetime reliability of FFs. Based on Eq. 11, TYR is computed for the FFs under study, taking into account a timing yield of 0.95 under a variation ratio of 15% for different operation time durations. Figure 6a, b show the computed TYR for the FFs (after 3, 6, and 9 years of operation, which is respectively shown as $\mathcal{R}_{0.95}(3)$, $\mathcal{R}_{0.95}(6)$, and $\mathcal{R}_{0.95}(9)$) for 0-to-1 and 1-to-0 output transitions, respectively. As expected, it can be observed that lifetime reliability decreases due to aging effects. In some cases, the lifetime reliability of FFs decreases up to 40% after 9 years of operation. Moreover, lifetime reliability decreases even more with increasing aging effects under higher operation time. The results also indicate that master-slave and pulsed FFs are not equally vulnerable to aging effects.

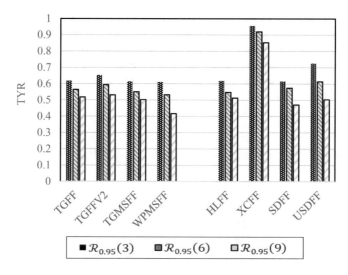

(a) TYR values considering 0-to-1 output transition

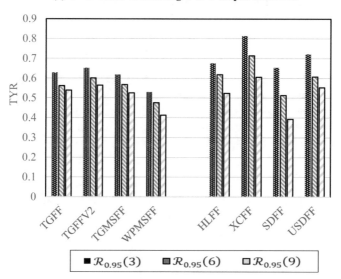

(b) TYR values considering 1-to-0 output transition

Fig. 6 Lifetime reliability (TYR) of the FFs considering different output transitions and after various operation times

As the results show, pulsed FFs are, overall, more reliable than master-slave FFs, considering the impacts of aging. Pulsed FFs include a clock pulse generation part in addition to the latching circuitry which generates narrow pulses as the clock signal. Hence, in this class of FFs, it is observed that many of the transistors are ON for a short duration of time in a clock period. As a result, the stress time of some transistors in pulsed FFs is less than that of master-slave FFs. Hence, it is observed that pulsed FFs are, overall, more reliable than master-slave ones. Among pulsed FFs, XCFF shows more reliability against aging effects. In XCFF, the clock pulse generator part is separate from the latching circuitry, causing the transistors in the XCFF circuits to less stressed. Hence, it is more robust to delay degradation induced by aging mechanisms.

5.4 Power-Delay-Product Comparison of FFs

In order to provide a more comprehensive analysis, different FF topologies are analyzed based on the power-delay-product (PDP) metric. Average PDP is calculated by multiplying the 95-percentile point of CDF (i.e., $CDF^{-1}(0.95)$) by the average power and delay (D2Q) of each FF. The delay and average power of the FFs for both fresh time (considering only process variation) and operation time (considering both process variation and aging effects) are calculated. The percentage of increase in PDP at the end of operation time (*AVINC*) is computed as:

$$AVINC = \frac{Aged - Designed}{Designed} * 100 \qquad (13)$$

where Designed and Aged respectively show the average PDP in designed and aged times. Table 3 shows the obtained results under a variation ratio of 15% and 3 years of operation time. The first column shows the FFs' names, and the second indicates the analysis (either fresh or operation time). The next columns respectively show the obtained average PDP and AVINC values. As it can be seen, the value of AVINC

Table 3 PDP comparison of FFs considering process variations and aging

AVINC (%)	Avg PDP(fJ)	State	FFs
TGFF	Fresh time	10.81	28.21
	Oper. time	13.86	
TGFFv2	Fresh time	13.89	26.66
	Oper. time	17.59	
TGMSFF	Fresh time	47.94	19.78
	Oper. time	57.42	
WPMSFF	Fresh time	30.61	50.68
	Oper. time	46.13	
HLFF	Fresh time	12.05	29.18
	Oper. time	15.56	
XCFF	Fresh time	26.24	0.09
	Oper. time	28.77	
SDFF	Fresh time	46.64	42.24
	Oper. time	66.35	
USDFF	Fresh time	25.85	22.20
	Oper. time	31.54	

increases in operation time up to 50% in write port master slave flip-flop (WPMSFF) whereas PDP is very low in FFs such cross charge control FF (XCFF).

6 Discussion and Conclusions

This chapter presents a comprehensive comparative analysis of the lifetime reliability of different FFs, considering the impact of process variations and aging mechanisms. Based on a novel proposed evaluation metric, TYR, the lifetime reliability of master-slave and pulsed FFs are compared, taking into account variations in transistor channel length and width, as well as NBTI and PBTI aging mechanisms. The reliability of both categories is measured under different variation ratios and operation lifetimes. According to the obtained results, the lifetime reliability decreases with increasing the aging effects in higher operation time and increasing the process variation ratios in future technology nodes.

Due to the presence of a clock pulse generation part in pulsed FFs, they show more reliability that master-slave ones against the aging impacts. However, among pulsed FFs, XCFF shows more reliability against aging effects as its clock pulse generator part is separate from the latching circuitry leading to less stressed transistors in the XCFF circuits. It is also observed that, that pulsed FFs are shows more robustness against PV effects comparing to that master-slave ones. However, among pulsed FFs, the reliability of HLFF decreases significantly as the variation ratio increases due to the lack of an internal feedback loop for state holding leading to more susceptibility to variability effects in output transitions and hence, more susceptibility to PV effects.

References

1. Alioto M, Consoli E, Palumbo G. Analysis and comparison in the energy-delay-area domain of nanometer CMOS Flip-Flops: Part I—methodology and design strategies. IEEE Trans Very Large Scale Integr (VLSI) Syst. 2011;19(5):725–36. https://doi.org/10.1109/TVLSI.2010.2041377.
2. Gupta MS, Rivers JA, Bose P, Wei G, Brooks D. Tribeca: design for PVT variations with local recovery and fine-grained adaptation. In: 2009 42nd annual IEEE/ACM international symposium on microarchitecture (MICRO), New York. 2009, p. 435–446. https://doi.org/10.1145/1669112.1669168
3. Schroder DK, Babcock JA. Negative bias temperature instability: road to cross in deep submicron silicon semiconductor manufacturing. J Appl Phys. 2003;94(1):1–18. https://doi.org/10.1063/1.1567461.
4. Asenov A, Kaya S, Brown AR. Intrinsic parameter fluctuations in decananometer MOSFETs introduced by gate line edge roughness. IEEE Trans Electron Devices. 2003;50(5):1254–60. https://doi.org/10.1109/TED.2003.813457.

5. Fukutome H, Momiyama Y, Kubo T, Tagawa Y, Aoyama T, Arimoto H. Direct evaluation of gate line edge roughness impact on extension profiles in sub-50-nm n-MOSFETs. IEEE Trans Electron Devices. 2006;53(11):2755–63. https://doi.org/10.1109/TED.2006.882784.
6. Mintarno E, Chandra V, Pietromonaco D, Aitken R, Dutton RW. Workload dependent NBTI and PBTI analysis for a sub-45nm commercial microprocessor. In: 2013 IEEE international reliability physics symposium (IRPS), Anaheim. 2013, p. 3A.1.1–3A.1.6. https://doi.org/10.1109/IRPS.2013.6531971
7. Alioto M, Consoli E, Palumbo G. Variations in nanometer CMOS flip-flops: part I – impact of process variations on timing. IEEE Trans Circuits Syst I Regular Papers. 2015;62(8):2035–43. https://doi.org/10.1109/TCSI.2014.2366811.
8. Alioto M, Consoli E, Palumbo G. Variations in nanometer CMOS flip-flops: part II—energy variability and impact of other sources of variations. IEEE Trans Circuits Syst I Regular Papers. 2015;62(3):835–43. https://doi.org/10.1109/TCSI.2014.2366813.
9. Jakushokas R, Popovich M, Mezhiba AV, Köse S, Friedman EG. Power distribution networks with on-chip decoupling capacitors. New York: Springer; 2011.
10. Rao VG, Mahmoodi H. Analysis of reliability of flip-flops under transistor aging effects in nano-scale CMOS technology. In: 2011 IEEE 29th international conference on computer design (ICCD), Amherst. 2011, p. 439–440. https://doi.org/10.1109/ICCD.2011.6081439
11. Golanbari MS, Kiamehr S, Tahoori MB, Nassif S. Analysis and optimization of flip-flops under process and runtime variations. In: Sixteenth international symposium on quality electronic design, Santa Clara. 2015, p. 191–196. https://doi.org/10.1109/ISQED.2015.7085423
12. Werner C, Backs B, Wirnshofer M, Schmitt-Landsiedel D. Resilience and yield of flip-flops in future CMOS technologies under process variations and aging. IET Circuits Devices Systems. 2014;8(1):19–26. https://doi.org/10.1049/iet-cds.2013.0122.
13. Khalid U, Mastrandrea A, Olivieri M. Effect of NBTI/PBTI aging and process variations on write failures in MOSFET and FinFET flip-flops. Microelectron Reliab. 2015;55(12 Part B):2614–26. https://doi.org/10.1016/j.microrel.2015.07.050.
14. Jafari A, Raji M, Ghavami B. Impacts of process variations and aging on lifetime reliability of flip-flops: a comparative analysis. IEEE Trans Device Mater Reliab. 2019;19(3):551–62. https://doi.org/10.1109/TDMR.2019.2933998.
15. Stojanovic V, Oklobdzija VG. Comparative analysis of master-slave latches and flip-flops for high-performance and low-power systems. IEEE J Solid State Circuits. 1999;34(4):536–48. https://doi.org/10.1109/4.753687.
16. Absel K, Manuel L, Kavitha RK. Low-power dual dynamic node pulsed hybrid flipflop featuring efficient embedded logic. IEEE Trans Very Large Scale Integr (VLSI) Syst. 2013;21(9):1693–704. https://doi.org/10.1109/TVLSI.2012.2213280.
17. Harris DM. Sequential element timing parameter definition considering clock uncertainty. IEEE Trans Very Large Scale Integr (VLSI) Syst. 2015;23(11):2705–8. https://doi.org/10.1109/TVLSI.2014.2364991.
18. Mahapatra S, et al. A comparative study of different physics-based NBTI models. IEEE Trans Electron Devices. 2013;60(3):901–16. https://doi.org/10.1109/TED.2013.2238237.
19. Kuhn KJ. Reducing variation in advanced logic technologies: approaches to process and design for manufacturability of nanoscale CMOS. In: 2007 IEEE International Electron Devices Meeting, Washington, DC. 2007, p. 471–474. https://doi.org/10.1109/IEDM.2007.4418976
20. Markovic D, Nikolic B, Brodersen RW. Analysis and design of low-energy flip-flops. In: ISLPED'01: proceedings of the 2001 international symposium on low power electronics and design (IEEE Cat. No.01TH8581), Huntington Beach, CA. 2001, p. 52–55. https://doi.org/10.1109/LPE.2001.945371
21. Papoulis A, Pillai SU. Probability, random variables and stochastic processes. London: McGraw-Hill; 2002.
22. Lanuzza M, De Rose R, Frustaci F, Perri S, Corsonello P. Impact of process variations on pulsed flip-flops: yield improving circuit-level techniques and comparative analysis. Proc PATMOS. 2011:180–9. https://doi.org/10.1007/978-3-642-17752-1_18.

23. Nanoscale Integration and Modeling (NIMO) Group. Predictive technology model. 2012. Available: http://ptm.asu.edu/latest.html
24. Tudor B et al. MOSRA: an efficient and versatile MOS aging modeling and reliability analysis solution for 45nm and below. In: 2010 10th IEEE international conference on solid-state and integrated circuit technology, Shanghai. 2010, p. 1645–1647. https://doi.org/10.1109/ICSICT.2010.5667399
25. Weste N, Eshraghian K. Principles of CMOSVLSI design, a system perspective. Reading: Addison-Wesley; 1993.
26. Zhuang N, Wu H. A new design of the CMOS full adder. IEEE J Solid-State Circuits. 1992;27: 840–4. https://doi.org/10.1109/4.133177.
27. Zimmermann R, Fichtner W. Low-power logic styles: CMOS versus pass-transistor logic. IEEE J Solid State Circuits. 1997;32:1079–90. https://doi.org/10.1109/4.597298.

Restructuring-Based Lifetime Reliability Improvement of Nanoscale Master-Slave Flip-Flops

1 Introduction

The scaling of technology to a nanometer regime leads to significant improvement in the performance and energy consumption of digital circuits. However, the functionality of digital circuits has been affected by different sources of variation in nanoscale integrated circuits [1, 2]. Manufacturing process variations (PVs), such as random dopant fluctuations (RDF) and line-edge roughness (LER), are considered the main source of variability in modern digital circuits [3, 4]. PV has led to considerable variations in circuit characteristics such as delay after manufacturing comparing to the designed circuit [4, 5]. Another source of variation aggravated in nanometer technology is transistor's aging in which the behavior of circuit is changed during the lifetime of the circuit. Aging mechanisms, especially bias temperature-induced instability (BTI), lead to increased threshold voltage in transistors. Hence, BTI increases the delay of gates and circuits, which, in turn, may lead to a violation of the timing constraints in circuits [6, 7]. Considering the impacts of PV and aging, it is indispensable to address these variation sources in reliable nanoscale digital circuit designs.

Flip-flops (FFs) are considered the key element in digital circuits, such as sequential circuits and pipeline stages. Hence, their functionality significantly affects the reliability of the whole chip. Moreover, the increasing larger variability due to PV and aging effects in nanometer regime continuously increases the probability of timing violations in nanoscale digital circuits systems including FFs [7, 8]. As a result, it is inevitable to consider PV and aging-induced variability during the reliability-aware design of FFs in modern very-large-scale integration (VLSI) circuits.

Some previous research studies have considered only PV effects in reliable FF designs [9, 10], while others introduce aging-aware techniques for a reliable FF design, ignoring PV effects [11, 12]. A transistor-sizing-based optimization method to reduce the effect of negative-bias temperature instability (NBTI) on the timing

© The Author(s), under exclusive license to Springer Nature Switzerland AG 2023
M. Raji, B. Ghavami, *Lifetime Reliability-aware Design of Integrated Circuits*,
https://doi.org/10.1007/978-3-031-15345-7_2

constraints of FFs is proposed in Ref. [11], taking into account static and near-static BTI stresses. This technique also reduces a large guardband cost, which is usually imposed when utilizing sizing-based improvement techniques. In Ref. [12], a low-cost strategy based on restructuring FFs is proposed to mitigate the impacts of aging on latches' robustness. The work only focuses on reducing soft error rate increase of FFs due to BTI effects and neglects FF timing reliability. In Ref. [12], the author strove to reduce the BTI effects on the soft error rate of latches and the critical charge of circuit nodes in latches. However, there is another challenge for designing a reliable latch/flip-flop regarding their timing constraints.

Although the mentioned previous works achieve valuable results, they address only one of the design challenges—i.e., the impacts of either PV or BTI. However, there are interdependencies between PV and BTI effects; i.e., the impact of aging mechanisms, such as BTI, depends on the amount of PV effects [13–16]. Hence, if design techniques ignore either PV or BTI effects, it will result in non-optimized impractical solutions for nanoscale digital circuits [13]. There are also some works that consider both PV and BTI effects for reliability analysis and/or the improvement of FFs. In Ref. [15], they present that independent consideration of the either process variation or aging may lead to significant inaccuracy compared to the combined effect analysis. Moreover, an optimization technique is proposed that increases the reliability of a given FF undergoing process variation and aging by changing the sizes of transistors. In Ref. [16], a transistor resizing technique is proposed to improve the resiliency of FFs against both PV and BTI effects. Although some works consider both PV and aging effects, most of them use sizing techniques that impose a large area to combat aging and process variation. Furthermore, none of them have focused on improving the timing reliability of FFs.

In this chapter, a lifetime reliability improvement technique is presented for the most popular group of FFs (i.e., master-slave flip-flops (MSFFs)) undergoing PV and aging effects [17]. In this technique, the internal structure of FFs is modified in order to increase FF lifetime reliability in the presence of process variation and BTI-induced variation. The stacked transistors in the feedback loop of the FF circuit are restructured in order to reduce their stress time (i.e., the time of being ON), thus improving the lifetime reliability of FFs. We apply our proposed timing yield-aware lifetime reliability (TYR) metric [7] as a fair metric for the evaluation of the lifetime reliability of FFs. Extensive experiments using Monte-Carlo HSPICE simulations are conducted to investigate the efficacy of the restructuring-based reliability improvement technique under various variation ratios and lifetime values. The experimental results show that the lifetime reliability of FFs improved by 16% at the expense of a 3% area overhead, 8% propagation delay absolute change, and 6% absolute power change in the presence of 20% variation ratio and 9 years of operation time. The results show that lifetime reliability improvement increases by increasing PV ratio and lifetime values.

The rest of the chapter is organized as follows: in Sect. 2, the procedure of the proposed technique for improving robustness against process variability is explained. We apply this strategy to selected circuits in each subsection. In Sect. 3, we measure lifetime reliability base on TYR metric under various conditions for both original and modified structures and present simulation results to verify the effectiveness of proposed technique. In Sect. 4, the costs of our strategy are evaluated, and finally, the chapter concludes in Sect. 5.

2   Proposed Lifetime Reliability Improvement Approach

We propose a technique to modify the internal structure of MSFFs with the aim of improving lifetime reliability without changing their functionality. In the following, we first describe the basic idea of the restructuring-based reliability improvement technique, and then we show how it is applied to other MSFF topologies.

2.1   Basic Idea

The basic idea behind our technique is to change the structure of transistors in the feedback loop of the MSFFs to reduce their stress time (i.e., the time of being ON), thus reducing the impact of BTI on such transistors during circuit lifetime. It is notable that BTI mainly affects the transistors that are ON (i.e., in a stress condition) for a long period of time [18]. Hence, BTI does not considerably influence the added pass transistors because they are constantly turned ON and OFF by the clock signal. So we ignore the negligible BTI effects on these transistors.

In the original circuits, we assume that each FF stores data with the same probability. Therefore, it is reasonable that the value of α in Eq. 2 is 0.5 for all n-channel metal-oxide semiconductor (NMOS) and p-channel metal-oxide-semiconductor (PMOS) transistors, which has an effect on the threshold voltage induced by BTI degradation. On the other hand, in the proposed technique, some transistors are conductive only half of the time compared to the original circuit. Since the transistors in the feedback loop do not need to turn on during the transparent part, the value of α for these transistors would reduce to 0.25.

The propagation delay of a gate is dependent on the value of the current strength of its pull-up/pull-down networks [19], which, in turn, is dependent on the threshold voltage of the transistors of those networks. Hence, by reducing BTI-induced threshold voltage variation, it leads to a reduction in the impacts of BTI on the current strength of the pull-up/pull-down network of the FF gates, causing a decrease in the impacts of BTI on its propagation delay and, therefore, FF reliability during the circuit's lifetime.

In the following subsections, we will show how the proposed technique is applied to some representative master-slave FFs in order to increase lifetime reliability.

2.2 Technique Application to TGFF

In order to describe the proposed technique, we begin by applying it to a transmission gate flip-flop (TGFF). The schematic structure of the considered FF is shown in Fig. 1a [20]. It includes two standard latches (i.e., master and slave latches) with similar structures and become active with complementary clocks. In the master latch, it is obvious that input D passes through the master to reach at slave part while clock is "0" and feedback loop is not conductive during this time. However, when the clock is "1," the pervious value of transparent stage is kept and the feedback loop is conductive. As can be seen in Fig. 1a, transistors P1 and N2 in the master part are driven by the output of inverter I1 independent of the clock. These two transistors are

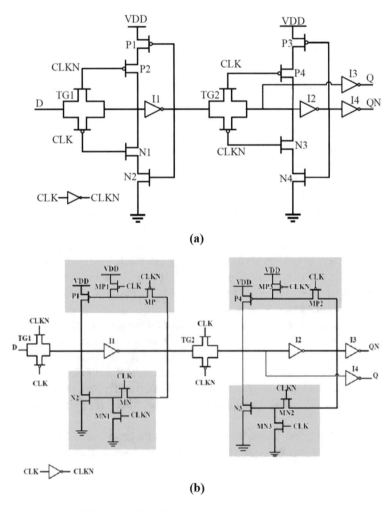

Fig. 1 Structure of TGFF and modified TGFF

turned on according to the value of output of I1 in both transparent and latch phase. Therefore, the probability of being in the stress condition leading to aging phenomenon for these transistors is 0.5 (a in Eq. 2). This trend is true for the slave part of transistors P3 and N4, which are in the feedback loop. We apply our method in these parts of the circuits to reduce the stress time ratio. The proposed modified FF is shown in Fig. 1b. In the master part of the modified TGFF, transistors P1 and N2 are driven by transistors MP and MN, respectively. During the latching phase in master part (when clock is "1"), P1 and N2 are driven through the pass transistors MP and MN and one of them is on according the value of output of the inverter. So similar to the original circuit, the feedback loop in this phase is conductive. Moreover, transistors MN1 and MP1 are both off during the latching phase. However, in the transparent phase, MP1 and MN1 are on to keep transistors P1 and N2 off. So they do not have to be on during the transparent phase, and the probability of these transistors to be in a stress condition decreases to 0.25 (a in Eq. 2). As a result, some transistors in the feedback loop in the modified circuit are conductive only during the latching phase and are off during the transparent phase, which reduces the stress time of these transistors.

2.3   Technique Application to TGFFV2

Figure 2a shows a schematic image of TGFFV2 [20]. Similar to TGFF, it is composed of master and slave latches. When the clock is "0," in the master latch, the input D propagates to the slave part and the feedback loop is not conductive while when the clock becomes "1," the value is stored by helping of feedback loop. Transistors P2 and N1 are driven by the output of inverter I1, and based on its value, either P2 or N1 is turned on during the transparent and latch phases. Therefore, the time ratio of being under stress is 0.5 (a in Eq. 2). The technique is applied to these transistors in order to reduce their stress time and keep them off during unnecessary time periods. A schematic diagram of the modified circuit is presented in Fig. 2b, where transistors P2 and N1 are driven by transistors MP and MN, respectively. In this case, during the latching phase (when the clock is "1"), transistors P2 and N1 are turned on based on the stored value through transistors MP and MN, and MP1 and MN1 are off in this phase. However, in the transparent phase (when the clock is "0"), transistors MN1 and MP1 are on. Consequently, both MP and MN are off during this time, and the probability of them being on declines to 0.25. Furthermore, similar to the master part, we modify the feedback loop of the slave part of the FF, as shown in Fig. 2b. As a result, transistors in the proposed circuit are conductive only during the latching phase instead of being on in all phases, and thus, the modified FF becomes more reliable against aging effects.

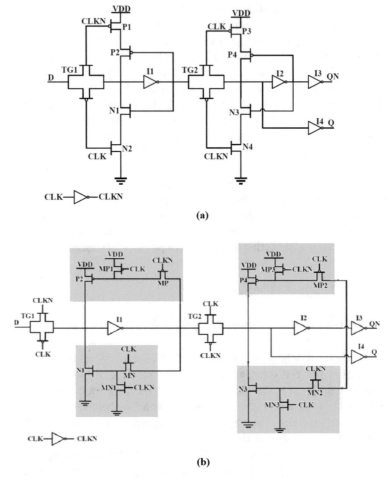

Fig. 2 Structure of TGFFV2 and modified TGFFV2

2.4   Technique Application to WPMS

The proposed technique is also applied to write port master-slave (WPMS) FFs, which are one of the most common master-slave circuits. A schematic image of WPMS is shown in Fig. 3a [21]. Similar to other MSFFs, when the clock is "0," the input propagates from the master latch to the slave latch, and the forward loop is not conductive in the master part. However, while the clock is "1," the master part keeps the previous value. The state of being on in the case of transistor N5 depends on the output of inverter I2 in the master latch. Similarly, N10 in the slave part is driven by inverter I4 and is turned on/off based on its output value. Therefore, these transistors are under extra stress independently of transparent and latch phases. Thus, the

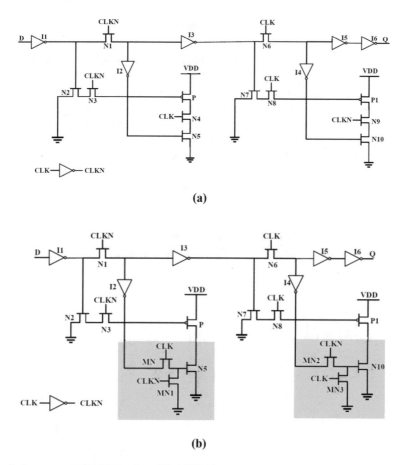

(a)

(b)

Fig. 3 Structure of WPMSFF and modified WPMS

proposed technique is aimed to decrease stress time by modifying the structure of the
FF circuit.

The modified circuit is presented in Fig. 3b. In this circuit, transistors N5 and N10
are driven by MN and MN2 transistors, respectively. Moreover, transistors N5 and
N10 are not on during the transparent phase, and, thus, the probability of being on is
decreased by half (from 0.5 to 0.25). Transistors MN1 and MN3 turn off transistors
N5 and N10 when they are not necessary to be on. This modification decreases the
stress time of transistors N5 and N10, and, therefore, the tolerance to aging of the
modified WPMS is increased, and its lifetime reliability is enhanced.

2.5   Technique Application to C2MOS

C2MOS is another master-slave FF, which we modify based on the proposed technique for increasing its lifetime reliability. The topology of C2MOS is presented in Fig. 4a [22]. It includes two latches with opposite clock phases. If the clock is "0," input D passes to the master latch and the slave latch holds the previous value. Otherwise, if the clock is "1," the input D is blocked from the master part and master part holds the value and passes value to slave. Each part (master or slave) contains one direct signal pass and one feedback loop, which are not active at the same time. Transistors P3 and N4 are driven by inverter I1 in the feedback loop of the master part when the clock is "1." However, they do not have to be on and under stress during the transparent phase. Therefore, the proposed technique is applied to the feedback loop containing these transistors.

The modified C2MOS is shown in Fig. 4b. As can be seen, transistors P3 and N4 are respectively driven by MP and MN during the latching phase in the master latch

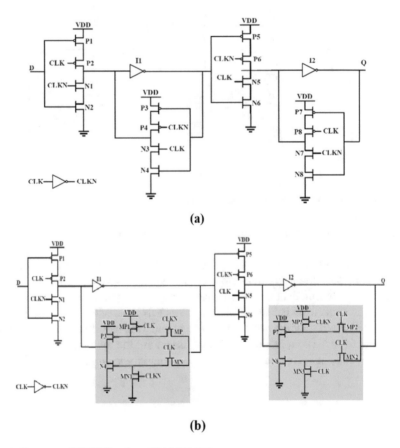

Fig. 4 Structure of C2MOS and modified C2MOS

(when the clock is "1"), and one of these transistors is turned on based on the stored value. Furthermore, both transistors MP1 and MN1 are off. However, when the clock becomes "0" (in the transparent phase), the feedback loop is not conductive and P3 and N4 are off by helping MP1 and MN1. Similarly, the proposed technique is applied to P7 and N8 in the slave latch. Consequently, in the modified FF, transistors are driven only during the latching phase, while they are off during the transparent phase in both master and slave parts. The modified structure reduces the stress time of some transistors in C2MOS by half. As a result, the lifetime reliability of the FF with a new structure is increased in comparison to that with the original structure.

2.6 Transistor Sizing

In this subsection, the details of transistor sizing of the modified FFs are reported considering 16-nm bulk CMOS predictive technology model (PTM) [23]. Table 1 reports the transistors aspect ratio (W/L) with $L = L_{MIN} = 16$ nm.

3 Experimental Results

In this section, we provide the details and results of the experiments conducted in order to show the efficacy of the proposed technique. In the following, we describe the simulation setup used in our experiments and then present the results of the characterization of the FFs under study. We show the impact of the proposed technique on delay values of FFs (i.e. D2Q values), present the lifetime reliability of FFs achieved by the proposed technique based on TYR metric presented in [14]. Finally, we investigate the effects of the proposed technique on other FF parameters. Also, the costs in terms of area, propagation delay, and the power consumption of both original and modified FFs are reported to provide a fair evaluation of the proposed technique.

3.1 Characterization Setup

The simulation setup for the FFs under study is presented in Fig. 5. Data and clock inputs are passed from two-stage buffers with minimum-sized inverters. The output of each circuit is loaded by a capacitor with a capacity of 2fF. This structure made it possible to have a realistic setup for measuring FF parameters.

The FFs under study are implemented using a 16-nm bulk CMOS predictive technology model (PTM) [23]. Simulation results, demonstrated in the following, are gained using HSPICE simulations. The threshold voltage shift (ΔV_{th}) induced by

Table 1 Transistor aspect ratio for our FF implementation

Aspect ratio	NMOS	PMOS
$W/L = 1$	NMOS of all transmission gates in the original and modified TGFF MN3 and MN1 in the modified TGFF NMOS of all transmission gates in the original and modified TGFFV2 MN1 and MN3 in the modified TGFFV2 NMOS in inverters I2, I4, and I6 in the original WPMS NMOS in inverter I2, I4, and I6 in the modified WPMS N1, N2, N4, N5, N6, N7, N9, and N10 in the original WPMS N1, N2, MN, MN1, N5, N6, N7, MN2, MN3, and N10 in the original WPMS NMOS of all inverters in the original and modified C2MOS	PMOS of all transmission gates in the original and modified TGFF PMOS of all transmission gates in the original and modified TGFFV2 P and P1 in the original and modified WPMS
$W/L = 2$	NMOS of all inverters in the modified and original TGFF N3 and N4 in the original TGFF N3 and MN2 in the modified TGFF NMOS of all inverters in the modified and original TGFFV2 N3 and N4 in the original TGFFV2 N3 and MN2 in the modified TGFFV2	MP1 and MP3 in the modified TGFF MP1 and MP3 in the modified TGFFV2 PMOS in inverters I2, I4, and I6 in the original WPMS PMOS in inverters I2, I4, and I6 in the modified WPMS PMOS of all modified and original C2MO PMOS of inverters in the modified and original C2MOS
$W/L = 3$	NMOS in inverters I1, I3, and I5 in the original and modified WPMS	–
$W/L = 4$	N1 and N2 in the original TGFF N2 and MN in the modified TGFF N1 and N2 in the original TGFFV2 N1 and MN in the modified TGFFV2 N8 in the original and modified WPMS	PMOS of all inverters in the original and modified TGFF PMOS of all inverters in the original and modified TGFFV2
$W/L = 6$	–	PMOS in inverters I1, I3, and I5 in the original WPMS PMOS in inverters I1, I3, and I5 in the modified WPMS

$W/L = 8$	–	P1 and P2 in the original TGFF P1 and MP in the modified TGFF P1 and P2 in the original TGFFV2 P2 and MP in the modified TGFFV2
$W/L = 12$	–	P3 and P4 in the original TGFF P4 and MP2 in the modified TGFF P3 and P4 in the original TGFFV2 MP2 and P4 in the modified TGFFV2
$W/L = 18$	N3 in the original and modified WMPS	–

Fig. 5 The simulation setup

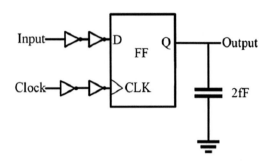

Table 2 Comparison of the timing parameters of the modified and original structures at design time

FFs	Transistor count		Transition	Setup time(ps)		Hold time(ps)		C2Q	
	Orig.	Mod.		Orig.	Mod.	Orig.	Mod.	Orig.	Mod.
TGFF	22	26	0-to-1	41.0	26.0	−20.5	−12.3	111.9	101.5
			1-to-0	27.0	20.4	−14.4	−10.5	119.1	113.9
TGFFV2	22	26	0-to-1	45.2	29.6	−12.7	−8.7	118.1	101.4
			1-to-0	20.1	8.5	−9.3	−5.4	128.8	114.2
WPMS	26	28	0-to-1	40.0	31.2	−16.3	−9.5	95.9	86.0
			1-to-0	140.0	119.6	−23.4	−17.6	69.8	57.7
C2MOS	20	24	0-to-1	79.0	74.5	−27.6	−18.2	46.5	44.7
			1-to-0	89.3	80.6	−19.6	−11.3	78.5	74.5

BTI degradation are evaluated using Eq. 2 in which the parameters are set based on BTI-induced delay degradation of a simple inverter becomes 10% over 3 years [26]. In order to find the parameters in BTI degradation, built-in HSPICE tool MOSFET model reliability analysis (MOSRA) tools [24] is used. In addition, the effects of PV on transistor width and length are determined using Eq. 3. Monte-Carlo-based HSPICE simulation is implemented, assuming normal distributions of transistor channel length and width. In order to do a statistical variability analysis, Monte-Carlo simulations are performed for 1000 iterations. The voltage and temperature are respectively considered 25 °C and 0.9 volts in nominal condition and clock period 3 ns. In these experiments, the impacts of process variations on transistor channel length and width are taken into consideration by modeling them as normally distributed random variability. The constants in Eqs. 1 and 2 are assumed as follows: $n = 1/6$, $a = 3.9 \times 10^{-3}$, and $A_0 = 10^{-9}$ [26], respectively.

Table 2 also compares the timing parameters of both original and modified circuits at design time. The parameter values are provided for both the 0-to-1 and 1-to-0 transitions of the FF output. Without any setup time violation, D2Q is defined as the propagation delay from clock edge to output. The setup time is defined as the time instance which leads to 10% increase in the C2Q delay with respect to the nominal C2Q delay (i.e., $1.1 \times C2Q_{min}$) [25].

3.2   Lifetime Reliability Increase

In this subsection, we take advantage of the TYR metric to compare the lifetime reliability of the modified structure and the original structure under different situations, considering the impacts of PV and BTI. The lifetime reliability (TYR in Eq. 4) of FFs for both states (original and modified) is calculated, and then the percentage of increase in TYR after restructuring (TYRINC) is computed as:

$$TYRINC(\%) = \frac{TYR_{original} - TYR_{modified}}{TYR_{original}} * 100 \tag{1}$$

where $TYR_{original}$ and $TYR_{modified}$ show the TYR of the original and modified FFs, respectively. Figure 6 shows the obtained results for the different experiment cases; i.e., TYR is computed for the FFs under study, taking into account a timing yield of 0.95 under variation ratios ($3\sigma/\mu$) of 10%, 20%, and 30% for different operation times (3, 6, and 9 years). It can be observed that the percentage of lifetime reliability increases with the increase in operation time; i.e., the efficacy of the proposed technique is increased with higher operation time. The amount of TYRINC increases in operation time up to 40% in TGFF after 9 years operation time. This can reach up to 25% under 30% process variation. The difference in the effectiveness of the proposed restructuring technique comes from the difference in the number of transistors under stress and their stress probability. The results show that the proposed technique achieves higher timing reliability improvement (i.e., TYRINC) in FFs with a higher number of transistors under stress and higher stress probabilities.

3.3   Cost Evaluation

We investigate the cost of the proposed restructuring technique in terms of area, input-output propagation delay, and average power. Equations 2, 3, and 4 show the method of calculating the cost of the modified structure relative to the original one for area (ΔArea), propagation delay (ΔPD), and power consumption (ΔPower), respectively. For both original and modified FFs, the total area is obtained by summing up the area of the transistors in the circuit. The area of an individual transistor is estimated as a function of the width and length of the transistor channel [27]. Moreover, in Eq. 2, ΔArea shows the percentage of area change, and $Area_{modified}$ and $Area_{original}$ respectively represent the area of the original and modified structures. A similar notation is used for the other cost parameters:

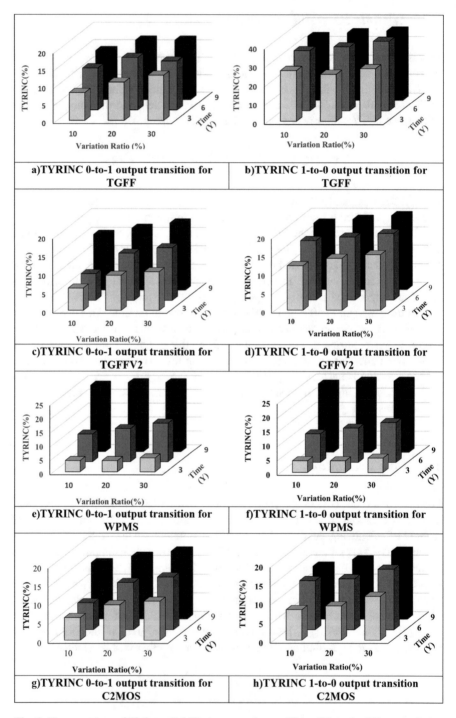

Fig. 6 The percentage of lifetime reliability increase after modifying FFs under different variation ratios and different years

Table 3 Cost comparison of the modified and original structures at design time

FFs	States	Area ($\times 10^{-3} \mu^2$)	(ΔArea) $\times$ 100	Propagation delay (ps)	(ΔPD) $\times$ 100	Power (nw)	ΔPower ($\times 100$)
TGFF	Orig.	20.4	7.5	149.5	-5.7	322.0	7.4
	Mod.	22.0		140.9		346.0	
TGFFV2	Orig.	20.4	7.5	155.9	-10.3	347.3	-0.3
	Mod.	22.0		139.8		346.0	
WPMS	Orig.	16.6	3.0	172.8	-7.8	409.7	20.5
	Mod.	17.1		159.2		493.7	
C2MOS	Orig.	76.8	18.8	146.5	-2.6	591.1	-2.1
	Mod.	91.2		142.6		578.6	

$$\Delta\text{Area}(\%) = \frac{\text{Area}_{\text{modified}} - \text{Area}_{\text{original}}}{\text{Area}_{\text{Original}}} * 100 \tag{2}$$

$$\Delta\text{PD}(\%) = \frac{\text{PD}_{\text{modified}} - \text{PD}_{\text{original}}}{\text{PD}_{\text{Original}}} * 100 \tag{3}$$

$$\Delta Power(\%) = \frac{\text{Power}_{\text{modified}} - \text{Power}_{\text{original}}}{\text{Power}_{\text{Original}}} * 100 \tag{4}$$

Table 3 represents the computed results for two structures at design time. As can be seen, the relative area increases for all selected FFs. This is because of the extra transistors added to the FFs in the restructuring technique. However, propagation delay may be decreased after restructuring. The proposed method adds some transistors to the feedback part of the FFs, and the forward path between the input and output is not modified. Therefore, it does not increase propagation delay and may even lead to a reduction in delay in the improved design. Propagation delay is calculated by computing the average value between both 0-to-1 and 1-to-0 output transition delays. Average power, which includes both static and dynamic power, is also measured. It is expected that the average power is increased in some modified FFs and decreased in others. The reason is that the additional transistors used in the approach are minimum-sized transistors. Therefore, the load capacitance of the clock signal increases, and thus, some extra parasitic capacitors emerged in the modified FFs. However, there are some transistors that are omitted, leading to less power consumption after applying the proposed technique.

4   Conclusion

The reliability of FFs has been challenged by the effects of PV and aging on nanoscale digital circuits. In this chapter, a restructuring-based technique is proposed to improve the timing reliability of master-slave FFs, mainly to address the degradation caused by PV and BTI effects. The proposed strategy modifies the internal structure of the FFs to increase their reliability by reducing the stress time of some transistors. Extensive Monte-Carlo HSPICE experiments are conducted to show the efficacy of the proposed approach in increasing the reliability of the FFs under different variation ratios and operation lifetimes, as well as the imposed costs in terms of area, propagation delay, and average power. The results show that, the proposed technique increases the reliability of FFs by 22% averagely in the extreme conditions (i.e., 20% variation ratio and 9 years of operation time) in expense of acceptable changes of other parameters (3% area overhead, 8% propagation delay absolute change, and 6% absolute power change).

References

1. Khan S, Hamdioui S. Trends and challenges of SRAM reliability in the nano-scale era. In: 5th international conference on design & technology of integrated systems in nanoscale era, Hammamet. 2010, p. 1–6. https://doi.org/10.1109/DTIS.2010.5487565
2. Ghavami B, Raji M, Rasaizadi R, Mashinchi M. Process variation-aware gate sizing with fuzzy geometric programming. Comput Electr Eng. 2019;78:259–70.
3. Leung G, Chui CO. Variability impact of random dopant fluctuation on nanoscale junctionless FinFETs. IEEE Electron Device Lett. 2012;33(6):767–9. https://doi.org/10.1109/LED.2012. 2191931.
4. Raji M, Ghavami B. Soft error rate reduction of combinational circuits using gate sizing in the presence of process variations. IEEE Trans Very Large Scale Integr (VLSI) Syst. 2017;25(1): 247–60. https://doi.org/10.1109/TVLSI.2016.2569562.
5. Ebrahimipour SM, Ghavami B, Raji M. Adjacency criticality: a simple yet effective metric for statistical timing yield optimisation of digital integrated circuits. IET Circuits Devices Syst. 2019;13(7):979–87.
6. Wang W, Yang S, Bhardwaj S, Vrudhula S, Liu F, Cao Y. The impact of NBTI effect on combinational circuit: modeling, simulation, and analysis. IEEE Trans Very Large Scale Integr (VLSI) Syst. 2010;18(2):173–83. https://doi.org/10.1109/TVLSI.2008.2008810.
7. Jafari A, Raji M, Ghavami B. Impacts of process variations and aging on lifetime reliability of flip-flops: a comparative analysis. IEEE Trans Device Mater Reliab. 2019;19(3):551–62. https://doi.org/10.1109/TDMR.2019.2933998.
8. Alioto M, Consoli E, Palumbo G. Variations in nanometer CMOS flip-flops: part I—Timing variations due to process variation. IEEE Trans Circuits Syst I Regular Papers. 2015;62(8): 2035–43. https://doi.org/10.1109/TCSI.2014.2366811.
9. Lanuzza M, et al. Comparative analysis of yield optimized pulsed flip-flops. Microelectron Reliab. 2012;52(8):1679–89. https://doi.org/10.1016/j.microrel.2012.03.024.
10. Rajaei R, Tabandeh M, Fazeli M. Low cost soft error hardened latch designs for nano-scale CMOS technology in presence of process variation. Microelectron Reliab. 2013;53(6):912–24. https://doi.org/10.1016/j.microrel.2013.02.012.

11. Golanbari MS, Kiamehr S, Ebrahimi M, Tahoori MB. Aging guardband reduction through selective flip-flop optimization. In: 2015 20th IEEE European Test Symposium (ETS), Cluj-Napoca. 2015, p. 1–6. https://doi.org/10.1109/ETS.2015.7138775

12. OmaÑa M, Edara T, Metra C. Low-cost strategy to mitigate the impact of aging on latches' robustness. IEEE Trans Emerg Top Comput. 2018;6(4):488–97. https://doi.org/10.1109/TETC.2016.2586380.

13. Lu Y, Shang L, Zhou H, Zhu H, Yang F, Zeng X. Statistical reliability analysis under process variation and aging effects. In: 2009 46th ACM/IEEE design automation conference, San Francisco. 2009, p. 514–9.

14. Basu S, Vemuri R. Process variation and NBTI tolerant standard cells to improve parametric yield and lifetime of ICs. In: IEEE Computer Society Annual Symposium on VLSI (ISVLSI'07), Porto Alegre. 2007, p. 291–298. https://doi.org/10.1109/ISVLSI.2007.85

15. Golanbari MS, Kiamehr S, Tahoori MB, Nassif S. Analysis and optimization of flip-flops under process and runtime variations. In: Sixteenth international symposium on quality electronic design, Santa Clara. 2015, p. 191–6. https://doi.org/10.1109/ISQED.2015.7085423

16. Golanbari MS, Kiamehr S, Ebrahimi M, Tahoori MB. Selective flip-flop optimization for reliable digital circuit design. In: IEEE transactions on computer-aided design of integrated circuits and systems. https://doi.org/10.1109/TCAD.2019.2917848

17. Jafari A, Raji M, Ghavami B. Timing reliability improvement of master-slave flip-flops in the presence of aging effects. IEEE Trans Circuits Syst I Regular Papers. 2020;67(12):4761–73. https://doi.org/10.1109/TCSI.2020.3024601.

18. Kim TT-H, Kong ZH. Impact analysis of NBTI/PBTI on SRAM VMIN and design techniques for improved SRAM VMIN. J Semicond Technol Sci. 2013;13(2):87–97.

19. Mahapatra NR, Tareen A, Garimella SV. Comparison and analysis of delay elements. In: The 2002 45th midwest symposium on circuits and systems, 2002. MWSCAS-2002, Tulsa. 2002, p. II–II. https://doi.org/10.1109/MWSCAS.2002.1186901

20. Markovic D, Nikolic B, Brodersen RW. Analysis and design of low-energy flip-flops. In: ISLPED'01: proceedings of the 2001 international symposium on low power electronics and design (IEEE Cat. No.01TH8581), Huntington Beach. 2001, p. 52–5. https://doi.org/10.1109/LPE.2001.945371

21. Consoli E, Palumbo G, Pennisi M. Reconsidering high-speed design criteria for transmission-gate-based master–slave flip-flops. IEEE Trans Very Large Scale Integr (VLSI) Syst. 2012;20(2):284–95. https://doi.org/10.1109/TVLSI.2010.2098426.

22. Suzuki Y, Hirasawa M, Odagawa K. Clocked CMOS calculator circuitry. In: 1973 IEEE international solid-state circuits conference. Digest of technical papers, Philadelphia. 1973, p. 58–9. https://doi.org/10.1109/ISSCC.1973.1155149

23. Nanoscale Integration and Modeling (NIMO) Group. Predictive technology model. 2012. Available: http://ptm.asu.edu/latest.html

24. Tudor B, et al. MOSRA: an efficient and versatile MOS aging modeling and reliability analysis solution for 45nm and below. In: 2010 10th IEEE international conference on solid-state and integrated circuit technology, Shanghai. 2010, p. 1645–7. https://doi.org/10.1109/ICSICT.2010.5667399

25. Kang S, Leblebici Y. CMOS digital integrated circuits. New York: McGraw-Hill; 2003.

26. Werner C, Backs B, Wirnshofer M, Schmitt-Landsiedel D. Resilience and yield of flip-flops in future CMOS technologies under process variations and aging. IET Circuits Devices Syst. 2014;8(1):19–26. https://doi.org/10.1049/iet-cds.2013.0122.

27. Weste N, Harris D. CMOS VLSI design: a circuits and systems perspective. New York: Addison-Wesley; 2004.

Lifetime Reliability Improvement of Pulsed Flip-Flops

1 Introduction

Although technology scaling into the nanometer regime has led to considerable improvement in the performance and energy consumption of digital circuits, modern digital systems have encountered serious new challenges. Manufacturing process variation (PV), such as random dopant fluctuation (RDF) and line edge roughness (LER), is a major issue in nanoscale integrated circuits. PV adversely affects various characteristics of digital circuits and systems after their fabrication [1–4]. This variation invalidates traditional design assessments, and hence, circuit designers have to take account of it in modern designs.

Transistor aging is a growing reliability issue in nanoscale CMOS technology. Bias temperature-induced instability (BTI) is an aging mechanism that increases the delay of gates over time, leading to timing constraint violations and, hence, reliability loss in digital circuits [5]. With the increase of PV in nanoscale technologies, BTI effects have also increased. So the impacts of both PV and BTI are necessary to consider to make reliable nanoscale digital circuit designs [6].

Pulsed FFs (PFFs) are known to be the most energy-efficient FFs in a large portion of the design space, ranging from high-speed designs to minimum energy-delay-product ones [7]. PFFs have two main parts: the latching part to store the input data and the pulsed generator part, which generates the pulse of the clock signal. Considering the impacts of PV and BTI, the timing characteristics of PFFs may be adversely affected over time, and thus, the lifetime reliability of nanoscale PFFs containing digital circuits may be considerably reduced.

There are a few previous works in which the reliability of FFs is improved, taking into account both BTI and the nanoscale design issue (i.e., PV). In Ref. [8], an optimization framework is proposed to mitigate the timing failure of FFs undergoing PV and BTI effects. This framework applies a transistor sizing technique to increase the FF's resilience against timing. However, this sizing-based approach imposes a large area and power overhead on the design and also neglects the timing reliability

© The Author(s), under exclusive license to Springer Nature Switzerland AG 2023
M. Raji, B. Ghavami, *Lifetime Reliability-aware Design of Integrated Circuits*,
https://doi.org/10.1007/978-3-031-15345-7_3

issue in FFs. Moreover, an independent consideration of the either process variation or aging, so it has an adverse effect on significant inaccuracy compared to the combined effect analysis. In Ref. [9], a restructuring technique is proposed to enhance the reliability of master-slave FFs undergoing process variation and aging effects. However, the proposed technique is only applicable in master-slave FFs.

In this chapter, a timing reliability improvement method is presented for PFFs, considering the impacts of BTI and PV [10]. The proposed method is based on the restructuring of a PFF circuit; i.e., the pulsed clocked signal in a pull-down network is modified, such that the stress time of some transistors is reduced and, thus, the reliability of the PFF is increased. In order to evaluate the reliability of PFFs, we use our previously introduced metric [6], called timing yield-aware lifetime reliability (TYR). The efficacy of the proposed method is investigated by conducting extensive experiments using Monte-Carlo HSPICE simulations under various variation ratios and lifetime values. According to the results of the experiment, the proposed method improves the lifetime reliability and area of FFs by 15% and 4%, respectively, while the propagation delay absolute change and absolute power change under a 30% process variation ratio and 9 years of operation time is 31% and 22%, respectively. The results also show that lifetime reliability improvement increases by increasing the PV ratio and lifetime values.

The rest of the chapter is organized as follows: Sect. 2 describes the proposed technique for improving the timing reliability of PFFs. In Sect. 3, the timing reliability of each PFF is evaluated using the TYR metric and by performing extensive experiments under various conditions. Moreover, in order to investigate the effectiveness of our strategy, timing reliability is measured for both original circuits and after modifying structures. The side effects of the proposed method are investigated in Sect. 4. Section 5 concludes the chapter.

2   Proposed Lifetime Improvement Approach

2.1   Basic Idea

BTI is highly dependent on the stress time of transistors; i.e., reducing stress time leads to reducing BTI effects. Based on this attribute, a technique is proposed to modify the structure of FFs to reduce the stress time on transistors. In this technique, the transistors are series with pulsed clock transistors in a pull-down network in order to decrease the time of being on (stress time). In the original circuit, it is assumed that the FFs store data with the same probability, so the value of stress time is set to 0.5 for the transistors in the FFs. However, the proposed technique causes some transistors to be conductive only when delayed pulses are applied. In addition, these transistors are not ON during the transparent phase; hence, the value of α for these transistors would decrease. As a result, reducing stress time (and, consequently, BTI impact on its propagation delay) improves the lifetime reliability of FFs. In the following subsections, our proposed technique is applied to some representative pulses, and then we explain this modification to improve lifetime reliability.

2.2   Application of the Technique to HLFF

We apply the proposed restructuring technique to hybrid latch flip-flops (HLFFs). A schematic diagram of this FF is demonstrated in Fig. 1a [10]. This circuit consists of a specific part to generate the clock pulse, and this clock would be applied to the circuit. In HLFF, the propagation delay of three inverters generates a delayed inverted clock signal to define the transparency window. There are two separate paths to write "0" and "1" in the output. Data can be transmitted to node Q if the clock and its delayed clock pulse are both high. Therefore, the FF is in a transparent phase. However, the FF is in a latch mode when node Q holds the previous value out of this transparency window. As shown in Fig. 1a, transistor N2 is driven by the value of input D. The probability of being in a stressed state (α) for N2, which leads to BTI for this transistor, is 0.5. Transistors N2 and N3 are series in the pull-down network. Therefore, both of these transistors should be simultaneously on to provide a conducting path in the pull-down network. To begin with, the proposed technique is applied in this part of the circuit to reduce the stress time ratio. Transistor N2 and N3 do not need to be series. The clock can be a control signal for the input of transistor N2 rather than input D. Figure 1b shows the modified structure for HLFF. According to the modified structure, transistor N1 is driven by transmission gate T1, which is controlled by the delayed clock signal. So the probability of a stressed condition decreases. This transistor does not have to be on during the transparent phase in spite of the original circuit. Moreover, this similar structure is also seen in the second part of the circuit among the transistors N5 and N6 in the original structure of HLFF. Therefore, we apply a similar technique to this part. As shown in Fig. 1b, transistor N3 in the modified circuit is driven by transmission gate T2 rather than node X in the original HLFF. Moreover, transistors M1 and M2 are both off during the latching phase.

Transistors M1 and M2 play a crucial role in keeping transistors N1 and N3 off during this unnecessary state. In addition, transistors N1 and N4 are both on by the clock signal in the original circuit. Consequently, it is possible to change these transistors into one, which is shared between the two parts. Transistor N2 in a modified structure is a shared transistor. These changes in the pull-down network do not affect the functionality of the circuit but reduce the stress time of transistors. Some transistors in the pull-down network do not have to be on during the latching phase, and hence, the probability of being in a stressed condition decreases. As a result, changing the FF structure leads to enhancement in the reliability of the whole circuit.

2.3   Application of the Technique to SDFF

The topology of a semi-dynamic flip-flop (SDFF) is illustrated in Fig. 2 [10]. It is one of the most famous PFF circuits. Similar to other PFFs, it has a pulse generator part,

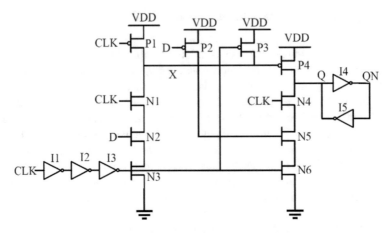

(a) The original structure of HLFF

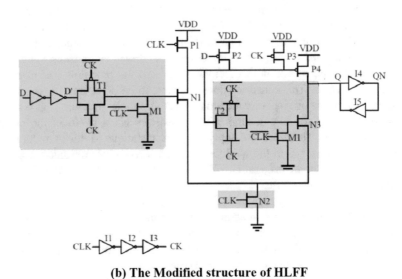

(b) The Modified structure of HLFF

Fig. 1 HLFF original structure and the modified HLFF based on the proposed restructuring technique

which provides a delayed clock signal to the circuit. The timing window is defined by two inverters I1, I2, and the NAND gate after the rising edge of the clock. In the original SDFF (Fig. 2a), transistors N1 and N2 are series in the pull-down network. The transistor N2 is always on based on the value of D. Therefore, the probability of being in the stress state (α) for N2 is 0.5. However, we change this part into Fig. 2b. Hence, this transistor is N1 in a modified circuit. Transistor N1 is driven by transmission gate T1 controlled by a pulsed clock. The role of transistor M1 in the modified SDFF is to turn transistor N1 off when the pulsed clock signal is not active.

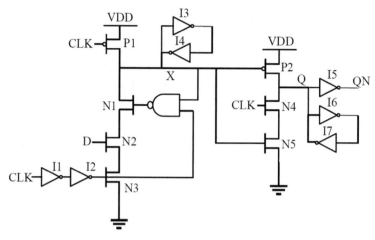

(a) The original structure of SDFF

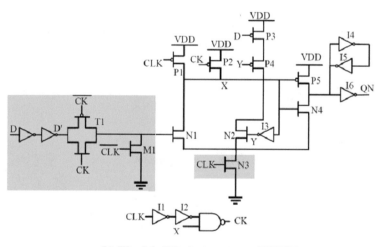

(b) The Modified structure of SDFF

Fig. 2 SDFF original structure and the modified SDFF based on the proposed restructuring technique

This method leads to stress reduction in the transistor. As a result, it turns on only when the transparent window is active. In addition, since transistors N3 and N4 in the original SDFF are driven by clock and delay clock, we combine them into one transistor. Transistor N2 is a common transistor between the first and second parts. So these changes contribute to decrease the stress time of transistor N1 compared to the original circuit. As a result, aging degradation is enhanced, which improves the lifetime reliability of SDFF.

2.4   Application of Technique to USDFF

Figure 3 demonstrates a schematic image of an ultraSPARC semi-dynamic flip-flop (USDFF) [10]. USDFF is also one of the most well-known PFFs. Inverters I1 and I2 and NAND produce a delayed clock signal to be applied to the circuit. In Fig. 3a, transistor N2 is turned on depending on the value of D. Therefore, the stress time

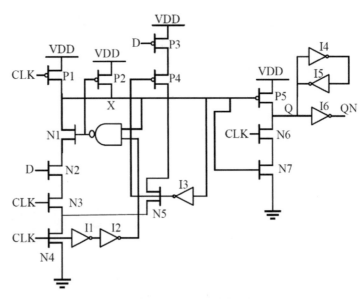

(a) The original structure of USDFF

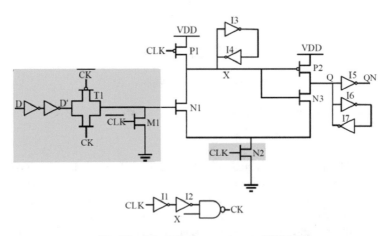

(b) The Modified structure of USDFF

Fig. 3 USDFF original structure and the modified USDFF based on the proposed restructuring technique

ratio is 0.5 (in Eq. 2). The restructuring method is applied to reduce the stress time of transistors and keep them off during unnecessary time periods. The schematic of the modified circuit is presented in Fig. 3b. In the modified USDFF, transistor N1 is driven by transmission gate T1, which is controlled by the pulsed clock. Therefore, this transistor is only conductive by the delayed clock rather than being on every clock cycle. Moreover, transistor M1 turns off transistor N1 when it is not necessary to be turned on. Since both transistors N3 and N6 are conducted by the clock signal in the original circuit, we combine these transistors into one (N3 plays this role of a shared transistor in Fig. 3b). Hence, we decrease the number of transistors, and the modified circuit becomes more reliable against aging-induced delay variations.

2.5 Technique Application to XCFF

Cross charge-control flip-flop (XCFF) [10] is another pulsed FF that is modified based on the proposed strategy. Figure 4a presents the topology of XCFF. Node X1 in XCFF is a dynamic node that is directly connected to the gate of N4 and drives two inverters: I1 and I2. X1 is connected to the gate of N3 in a pull-down network. In the original circuit, transistors N2 and N3 are in series. They are conducted by D and the output of inverter I2, respectively. The stress time ratio for transistor N2 is 0.5 (Eq. 2). Therefore, we change the structure of this part of the circuit to reduce stress time. The modified XCFF is shown in Fig. 4b. As can be seen, transistor N2 is driven by transmission gate T1. So transistor N2 in the modified circuit is turned on when node X1 is active. Transistor M1 is used to keep transistor N2 off during an unnecessary state. The modified structure reduces the stress time of some transistors in XCFF decrease. The lifetime reliability of the FF with the new structure is increased as compared to that with the original structure.

3 Experimental Results

In this section, the efficacy of our strategy is investigated through extensive experiments. In the following section, the simulation setup for this experiment is described, and then a comparison of the lifetime reliability of both original and modified circuits under different years is shown. Moreover, lifetime reliability is studied under different process variations and lifetimes based on the TYR metric proposed in Ref. [6]. Finally, other parameters of FFs, such as delay and costs in terms of area, propagation delay, and the power consumption of both original and modified FFs, are investigated.

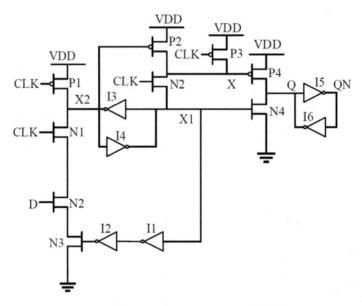

(a) The original structure of XCFF

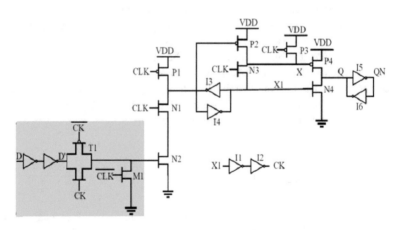

(b) The Modified structure of XCFF

Fig. 4 XCFF original structure and the modified XCFF based on the restructuring technique

3.1 Characterization Setup

The FFs are implemented and simulated in HSPICE using the predictive technology model (PTM) library 16-nm LP [17]. The temperature and supply voltage are 25 °C and 0.9 volts for nominal conditions, respectively. According to Eq. 2, the threshold voltage shift (ΔV_{th}) due to the effects of PV effects on width and lengths are calculated. Moreover, we use the MOSFET model reliability analysis (MOSRA)

Fig. 5 The simulation setup

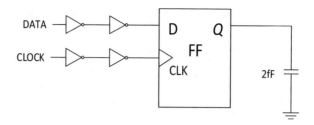

Table 1 The impacts of the proposed restructuring technique on other design parameters (area, delay, power consumption)

FFs	State	Transistor count	Propagation delay (Ps)	ΔPD (×100)	Average power(μw)	ΔPower (×100)
HLFF	Orig.	20	63.12	–	0.45	35.55
	Mod.	27	55.41	12.21	0.61	
SDFF	Orig.	25	53.89	–	2.31	14.71
	Mod.	30	33.65	37.55	2.65	
USDFF	Orig.	28	122.76	−8.76	1.26	29.36
	Mod.	32	112.87		1.63	
XCFF	Orig.	21	72.23	–	0.64	6.25
	Mod.	27	64.92	10.12	0.68	
CPFF	Orig.	23	64.42	–	0.52	36.53
	Mod.	31	52.12	19.09	0.71	

[18] built-in HSPICE tool to have NBTI and PBTI analysis. Process variations impact is investigated using HSPICE Monte Carlo simulations with 1000 runs by assuming normal distributions of the transistor channel length and width. To have a realistic running condition so as to evaluate parameter, Fig. 5. sow the simulation setup for each FF, which consist of the two-stage buffer with minimum sized inverters are used for data and clock inputs. A capacitor with capacity 2fF is also used as the load in the output.

3.2 FF Characterization Results

The characteristics of our selected pulsed FF at design time without the presence of PV and BTI are demonstrated in Table 1. To have a fair comparison, conventional pulsed FFs (CPFFs) [10] are also added in this table. As can be seen, some timing parameters that play a vital role in the function of FFs are measured. Parameters such as $C2Q_{min}$, setup time, D2Qmin. They are also calculated for both 0-to-1 and 1-to-0 transitions FF's output. The C2Q is defined as the delay from the clock edge to output. the setup time is defined as the time instance which leads to 10% increase in the C2Q delay with respect to the nominal C2Q delay and D2Q is the summation of C2Q and setup time.

3.3   Lifetime Reliability of Both Structures

In this section, we investigate the impact of the restructuring technique on lifetime reliability. So the lifetime reliability of both original and modified circuits is calculated. Figure 6 shows these amounts for all considered pulsed FFs. Based on TYR metric (Eq. 5), the lifetime reliability for both original and modified structure is measured. The results are obtained for timing yield ($\mathcal{Y}$) of 0.95 and 3, 6, and 9 years of operation time under process variations (with variation ratio ($3\sigma/\mu$) of 10%).

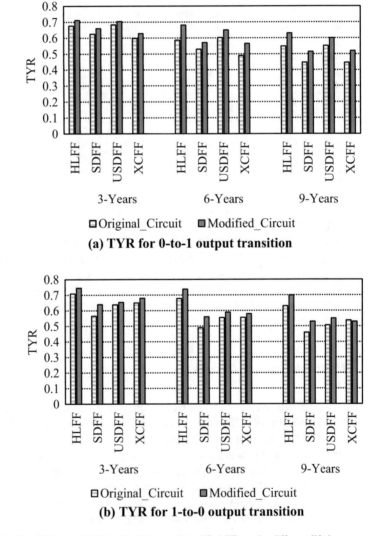

(a) TYR for 0-to-1 output transition

(b) TYR for 1-to-0 output transition

Fig. 6 The lifetime reliability of original and modified FFs under different lifetimes

Figure 6 shows the TYR for both 0-to-1 and 1-to -0 output transitions. In each figure, the black and grey bar chart shows the TYR of the modified and original circuits, respectively. Consider HLFF as an example; in this circuit, the TYR increases from around 0.58 to 0.68 in the 0-to-1 output transition. As a result, the lifetime reliability of the modified structure in all FFs for all operation time improves, and the gap between the original and modified structures increases with an increase in operation time.

3.4 Lifetime Reliability Increase

In this section, we compare the lifetime reliability of these two structures using the TYR metric under different process variation ratios and operation times. Lifetime reliability is calculated using Eq. 5 for the original and modified FFs, and then the percentage of increased TYR after applying the technique (TYRINC) is computed as:

$$\text{TYRINC} = \frac{\text{TYR}_{\text{original}} - \text{TYR}_{\text{modified}}}{\text{TYR}_{\text{original}}} * 100 \tag{1}$$

where $\text{TYR}_{\text{original}}$ and $\text{TYR}_{\text{modified}}$ indicate the TYR of the original and modified FFs, respectively. Figure 7 illustrates the percentage of increase in lifetime reliability for all pulsed FFs in both 0-to-1 and 1-to-0 output transitions. The results are obtained by considering a timing yield of 0.95 under variation ratios (VR) ($3\sigma/\mu$) of 10%, 20%, and 30% at various operation times (3, 6, and 9 years). As can be seen, the percentage of reliability increase is increased with higher operation times and process variation ratios. For instance, lifetime reliability improves by 25% in SDFF under a 30% process variation ratio and 9 years of operation time, which indicates the efficacy of the proposed restructuring method.

3.5 Cost Evaluation

In this section, the side effect of the proposed restructuring technique on other design parameters (transistor count, input-output propagation delay, and power consumption). In Eqs. 2 and 3 show the computation formula of the propagation delay change (which is calculated by measuring the average value between both 0-to-1 and 1-to-0 output transition delays) (ΔPD), and the power consumption change (ΔPower) of the proposed technique. As a result, the cost of the modified structure relative to the original one for these parameters of FFs can be calculated:

$$\Delta\text{PD}(\%) = \frac{\text{PD}_{\text{modified}} - \text{PD}_{\text{original}}}{\text{PD}_{\text{Original}}} * 100 \tag{2}$$

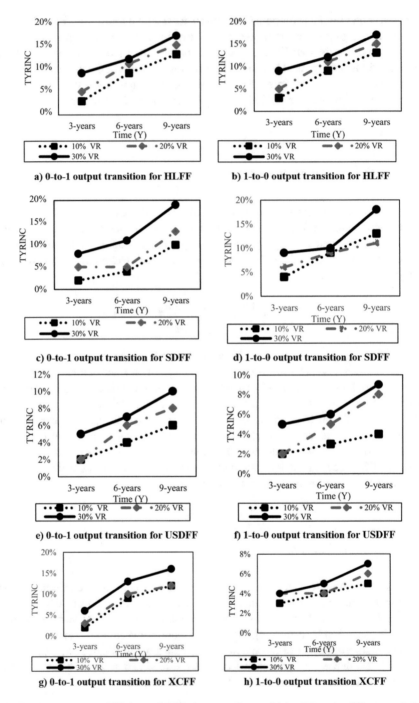

Fig. 7 The percentage of lifetime reliability increase after modifying FFs under different variation ratios and different years

$$\Delta Power(\%) = \frac{Power_{modified} - Power_{original}}{Power_{Original}} * 100 \qquad (3)$$

The side effect of the proposed restructuring technique on other design parameters (area, propagation delay, and power consumption) for both original and modified structures of PFFs are reported in Table 1. The delay decreases by restructuring the circuit while the trend of average power changes increases for modified PFFs. The reason of increasing power consumption is that the load capacitance of the clock signal increases due to added minimum sized transistors by applying this technique and thus, some extra parasitic capacitors have emerged in the modified FFs.

4 Conclusion

This chapter presents a restructuring-based technique to enhance the timing reliability of PFFs, considering PV and BTI effects, in nanoscale digital circuits. In this method, the internal structure of FFs is modified such that the stress time of important and effective transistors is reduced. Extensive Monte-Carlo HSPICE experiments are conducted to investigate the effectiveness of the proposed technique on the timing reliability of PFFs under different process variation ratios and operational times. According to the results obtained, this proposed method can raise the reliability of FFs by 15% at the expense of 4% area change under 30% process variation and 9-year operational time.

References

1. Alioto M, Consoli E, Palumbo G. Analysis and comparison in the energy-delay-area domain of nanometer CMOS flip-flops: part I—methodology and design strategies. IEEE Trans Very Large Scale Integr (VLSI) Syst. 2011;19(5):725–36.
2. Ghavami B, Raji M, Rasaizadi R, Mashinchi M. Process variation-aware gate sizing with fuzzy geometric programming. Comput Electr Eng. 2019;78:259–70.
3. Raji M, Ghavami B. Soft error rate reduction of combinational circuits using gate sizing in the presence of process variations. IEEE Trans Very Large Scale Integr (VLSI) Syst. 2017;25(1): 247–60. https://doi.org/10.1109/TVLSI.2016.2569562.
4. Ebrahimipour SM, Ghavami B, Raji M. Adjacency criticality: a simple yet effective metric for statistical timing yield optimisation of digital integrated circuits. IET Circuits Devices Syst. 2019;13(7):979–87.
5. Raji M, Mahmoudi R, Ghavami B, Keshavarzi S. Lifetime reliability improvement of nanoscale digital circuits using dual threshold voltage assignment. IEEE Access. 2021;9:114120–34. https://doi.org/10.1109/ACCESS.2021.3103200.
6. Jafari A, Raji M, Ghavami B. Impacts of process variations and aging on lifetime reliability of flip-flops: a comparative analysis. IEEE Trans Device Mater Reliab. https://doi.org/10.1109/TDMR.2019.2933998

7. Lin J. Low-power pulse-triggered flip-flop design based on a signal feed-through. IEEE Trans Very Large Scale Integr (VLSI) Syst. 2014;22(1):181–5. https://doi.org/10.1109/TVLSI.2012. 2232684.
8. Golanbari MS, Kiamehr S, Tahoori MB, Nassif S. Analysis and optimization of flip-flops under process and runtime variations. In: Sixteenth international symposium on quality electronic design, Santa Clara. 2015, p. 191–6. https://doi.org/10.1109/ISQED.2015.7085423
9. Jafari A, Raji M, Ghavami B. Timing reliability improvement of master-slave flip-flops in the presence of aging effects. IEEE Trans Circuits Syst I Regular Papers. https://doi.org/10.1109/ TCSI.2020.3024601
10. Jafari A, Raji M, Ghavami B. BTI-aware timing reliability improvement of pulsed flip-flops in nano-scale CMOS technology. IEEE Trans Device Mater Reliab. 2021;21(3):379–88. https:// doi.org/10.1109/TDMR.2021.3102521.

Gate Sizing-Based Lifetime Reliability Improvement of Integrated Circuits

1 Introduction

The key sources of failures that may threaten the reliability of digital circuits can be divided into time-independent failures, like process variations (PVs), and time-dependent failures, like the negative-bias temperature instability (NBTI) aging phenomenon.

Process variation is recognized as an important issue that affects integrated circuits' reliability. As fabrication technology continues to scale down, fabrication-induced parameter variations, such as changes in gate oxide thickness and threshold voltage, will significantly affect circuit timing and performance. Fabrication-induced PV may change circuit frequency by up to 30%. In order to meet specific timing constraints and increase circuit reliability, PV-induced uncertainty should be taken into consideration during the design phase.

Circuit performance not only changes with static PV but also degrades with time due to aging effects, resulting in the chips suffering from reliability problems during their lifetime. NBTI occurs when a p-channel metal-oxide-semiconductor (PMOS) transistor is negatively biased and manifests itself as an increase in transistor threshold voltage, which, in turn, causes degradation of device delay; i.e., device delay is increased. Since NBTI strongly affects circuit performance during its lifetime, it may lead to a violation of timing constraints. Because of NBTI effects, an increase in circuit delay for 65-nm technology is up to 30% [1]. In order to meet the specific timing constraint and increase the reliability of the circuit, design guidelines for NBTI-tolerant logic designs are necessary during the design phase.

Since, on one hand, NBTI-induced threshold voltage shift in a transistor depends on both environmental variations (like temperature and input signal probability) and transistor parameters (such as threshold voltage and gate oxide thickness) and, on the other hand, these parameters may change due to process variations, NBTI and PV effects have a strong impact on each other; i.e., they are significantly interdependent

M. Raji, B. Ghavami, *Lifetime Reliability-aware Design of Integrated Circuits*,
https://doi.org/10.1007/978-3-031-15345-7_4

[10]. As a result, the joint effects of PV and NBTI should be taken into account in the design and optimization flow of digital circuits.

In this chapter, we introduce a statistical framework based on incremental gate sizing to improve circuit lifetime reliability while considering the joint effects of PV and transistor aging. To this end, we introduce a canonical first-order delay model so that we are able to estimate the delay degradation of a gate under NBTI and PV. Using the proposed gate delay model, we present a statistical static timing analysis (SSTA) method, based on which we can estimate the performance of the circuit, taking into account the joint effects of NBTI and PV and considering spatial correlations. In order to design a circuit that meets timing constraints, an incremental gate sizing procedure is introduced. In this procedure, we first compute the *criticality* of each gate, which is defined as the probability that a gate lies on the critical path, taking into consideration NBTI and PV effects. Then we rank the gates based on their criticality, and finally, in order to apply gate sizing efficiently, a group of gates with the highest ranking is chosen for optimization. It is worthy to note that by using the proposed statistical gate delay model, we can compute the criticality of each gate incrementally, leading to significant speedup in the optimization flow.

2 Proposed Framework

Figure 1 shows a general view of the proposed framework to optimize the reliability of a circuit, considering the joint effects of NBTI and PV. The framework consists of three main parts:

1. *Statistical gate delay model*: by merging the uncertainty caused by PV effects into the NBTI model, a statistical gate-level NBTI- and PV-aware model is obtained. Using this model, we can characterize the PV and NBTI effects for each gate delay.
2. *Statistical circuit-level delay computation*: using the statistical gate delay model, block-based SSTA is performed to compute the statistical distribution of circuit delay. In order to compute delay distribution, a directed acyclic graph of the circuit is created. Then using SSTA and the runtime conditions of the circuit (such as lifetime and temperature), we can compute the lifetime delay of the circuit.
3. *Gate sizing algorithm*: in this part, the criticality of each gate is calculated to measure its impact on the circuit's lifetime. Then a group of gates with the largest criticality is selected to be used in the gate sizing procedure for optimizing circuit reliability. This procedure will continue until lifetime timing constraints are met or the area constraint is violated.

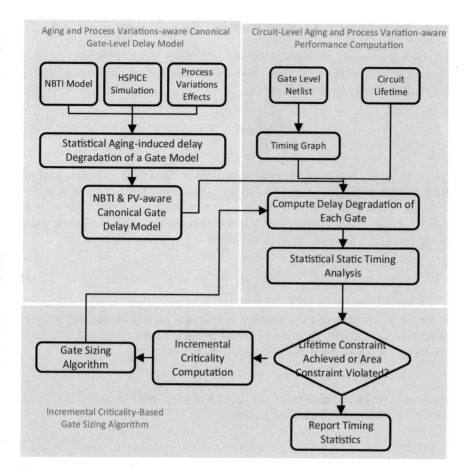

Fig. 1 Proposed framework

2.1   Statistical Gate Delay Model Under the Joint Effects of NBTI and PV

The delay of a gate at time t, considering the joint effects of NBTI and PV, can be expressed as follows:

$$D_{gate_i}(PV, t) = D_{gate_i}(PV, 0) + \Delta D_{gate_i}(PV, t) \qquad (1)$$

where $D_{gate_i}(PV, 0)$ is the initial gate delay after chip fabrication ($t = 0$) under PV effects and $\Delta D_{gate_i}(PV, t)$ represents aging- and PV-induced lifetime delay degradations.

2.1.1   Initial Gate Delay Under PV Effects

We applied a widely used canonical first-order delay model with relatively low complexity. The delay of a gate under PV effects, taking into account spatial correlation, can be stated as follows [3]:

$$D_{gate_i} = d_0 + \sum_{i=1}^{n} d_i \Delta V_{th(i)} + d_{n+1} \Delta V_{th\_rand} \tag{2}$$

where $\Delta V_{th(i)}$ represents an independent random variable used to express the spatially correlated threshold variations, ΔV_{th_rand} represents the residual independent variation of threshold voltage, d_i and d_{n+1} represent the sensitivity of the delay to each of the random variables, and d_0 is the mean delay, which can be calculated using the Alpha power law delay model [4]:

$$D = \frac{C_L V_{dd}}{\beta (V_{dd} - V_{th})^\alpha} \tag{3}$$

where D is the gate delay; C_L and α represent load capacitance and velocity saturation, respectively; V_{th} is the threshold voltage; β is a parameter that depends on the size of the gate; and V_{dd} is the power supply voltage.

2.1.2   Delay Degradation Under the Joint Effects of NBTI and PV Considering Spatial Correlation

In order to estimate the delay degradation of a PMOS transistor, the widely accepted predictive model in Ref. [26], which is a model based on a reaction-diffusion mechanism is used; that is:

$$\Delta V_{th\_nbti} = \left(\frac{\sqrt{K_v^2 \, s \, T_{clk}}}{1 - \beta_t^{\frac{1}{2n}}} \right)^{2n} \tag{4}$$

$$\beta_t = 1 - \left(\frac{2\varepsilon_1 t_e + \sqrt{\varepsilon_2 \, C \, (1-s) T_{clk}}}{2t_{ox} + \sqrt{Ct}} \right) \tag{5}$$

$$K_v = \left(\frac{q t_{ox}}{\varepsilon_{ox}} \right)^3 K_1^2 \, C_{ox} (V_{gs} - V_{th}) \sqrt{C} \, e^{\left(\frac{2E_{ox}}{E_{o1}} \right)} \tag{6}$$

where K_v is a function of temperature, electrical field, and carrier concentration; β_t is the fraction parameter of recovery; t represents transistor lifetime; V_{th} is the initial threshold voltage of the transistor; T_{clk} and s are the clock cycle time and signal transition probability (duty cycle), respectively; and n is determined by a dispersion material, which is around 0.16 for H_2. For a more detailed explanation of the model parameters, please refer to Ref. [5].

By incorporating PV-induced uncertainty into the conventional NBTI model, considering quad-tree spatial correlation [6], we can estimate threshold voltage degradation due to NBTI and PV [2]:

$$\Delta V_{th\_nbti} = A\left(1 - \gamma\left(\sum_{i=1}^{n}\Delta V_{th(i)} + \Delta V_{th\_rand}\right)\right)s^n t^n \qquad (7)$$

where A is a fitting parameter related to a specific process and operational conditions and γ denotes the sensitivity of NBTI-induced degradation to PV effects, which can be calculated by HSPICE simulations.

In order to compute the delay degradation of a gate due to NBTI and PV, we transform Eq. 7 into a canonical form as follows:

$$\Delta V_{th\_nbti} = As^n t^n + \sum_{i=1}^{n}(-As^n t^n \gamma)_i \Delta V_{th(i)} + (-As^n t^n \gamma)_{n+1}\Delta V_{th\_rand} \qquad (8)$$

Based on the Alpha power law delay model [4], the delay degradation of a gate can be modeled as a linear function of the threshold voltage shift; that is:

$$\Delta D_{gate_i}(PV, t) = \mu\Delta V_{th}(PV, t) \qquad (9)$$

$$\mu = \frac{\alpha D}{V_{dd} - V_{th0}} \qquad (10)$$

Therefore, by substituting Eq. 8 with Eq. 9, the delay degradation of a gate, taking into account the joint effects of NBTI and PV, can be expressed as follows:

$$\Delta D_{gate_i} = \mu As^n t^n + \sum_{i=1}^{n}(-\mu As^n t^n \gamma)_i \Delta V_{th(i)} + (-\mu As^n t^n \gamma)_{n+1}\Delta V_{th\_rand} \qquad (11)$$

2.2 Statistical Circuit-Level Delay Computation Considering the Joint Effects of NBTI and PV

Before explaining the proposed SSTA, some parameters must be introduced:

1. *Required arrival time (RAT)*: it pertains to "the time interval during which the data must arrive at some internal node" [7].

2. *Arrival time (AT)*: it is "the time at which data arrives at some internal node. It incorporates all the net and logic delays between the reference input point and the destination node" [7].
3. *Source node:* it is a virtual node connected to all primary inputs.
4. *Sink node:* it is a virtual node connected to all primary outputs or timing test nodes.

Given the circuit netlist and the NBTI- and PV-aware statistical gate delay model as inputs, the proposed NBTI- and PV-aware block-based SSTA method computes the effects of NBTI and PV of each logic gate, and carries out circuit-level statistical timing analysis with linear runtime complexity [7]. Generally, SSTA consists of two repeating steps:

1. *Signal propagation*: it refers to a condition in which a signal propagates from a gate input to the gate output.
2. *Merging arrival times*: this step merges multiple gate input arrival times to condense them to a gate output arrival time. The result of the condensation is the sum of the maximum input arrival times and gate delay. Consider the example shown in Fig. 2. Let $A_1 = a_0 + \sum_{i=1}^{n} a_i \Delta V_{th(i)} + a_{n+1} \Delta V_{th\_rand}$ and $A_2 = b_0 + \sum_{i=1}^{n} b_i \Delta V_{th(i)} + b_{n+1} \Delta V_{th\_rand}$ be the arrival times at node C and $D = d_0 + \sum_{i=1}^{n} d_i \Delta V_{th(i)} + d_{n+1} \Delta V_{th\_rand}$ be the delay of this node. The arrival time at G is expressed as:

$$
\begin{aligned}
G &= max\,(A_1, A_2) + C \\
&= max\left(a_0 + \sum_{i=1}^{n} a_i \Delta V_{th(i)} + a_{n+1} \Delta V_{th_{rand}},\ b_0 + \sum_{i=1}^{n} b_i \Delta V_{th(i)} \right. \\
&\qquad \left. + b_{n+1} \Delta V_{th\_rand}\right) + d_0 + \sum_{i=1}^{n} d_i \Delta V_{th(i)} + d_{n+1} \Delta V_{th\_rand} \\
&= e_0 + \sum_{i=1}^{n} e_i \Delta V_{th(i)} + e_{n+1} \Delta V_{th_{rand}} + d_0 + \sum_{i=1}^{n} d_i \Delta V_{th(i)} + d_{n+1} \Delta V_{th\_rand} \\
&= g_0 + \sum_{i=1}^{n} g_i \Delta V_{th(i)} + g_{n+1} \Delta V_{th\_rand}
\end{aligned}
\tag{12}
$$

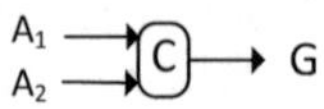

Fig. 2 Part of a timing graph

2.2.1   Arrival Time Propagation

In this procedure, gate delay is added to arrival time. If A_{gate_i}, D_{gate_i}, and ΔD_{gate_i} respectively represent the arrival time, delay, and delay degradation of a gate in canonical form, we have:

$$A_{gate_i} = a_0 + \sum_{i=1}^{n} a_i \Delta V_{th(i)} + a_{n+1} \Delta V_{th\_rand} \tag{13}$$

$$D_{gate_i} = d_0 + \sum_{i=1}^{n} d_i \Delta V_{th(i)} + d_{n+1} \Delta V_{th\_rand} \tag{14}$$

$$\Delta D_{gate_i} = \xi_0 - \sum_{i=1}^{n} \xi_i \Delta V_{th(i)} - \xi_{n+1} \Delta V_{th\_rand} \tag{15}$$

The sum operation can be computed as follows:

$$\begin{aligned}
X &= x_0 + \sum_{i=1}^{n} x_i \Delta V_{th(i)} + x_{n+1} \Delta V_{th\_rand} \\
x_0 &= a_0 + \xi_0 + d_0 \\
x_i &= (a_i - \xi_i) + d_i \\
x_{n+1} &= \sqrt{(a_{n+1} - \xi_{n+1})^2 + d_{n+1}^2}
\end{aligned} \tag{16}$$

2.2.2   Merging Arrival Times

In this step, we should compute the maximum arrival time of a gate's inputs. Due to the nonlinearity of the maximum operation, the maximum of two parameters in the canonical form cannot be expressed in the canonical form. In Ref. [7], an algorithm is presented for computing the statistical approximation of the maximum of two arrival times: A and B. First, it computes the variance and covariance of A and B as:

$$\sigma_A^2 = \sum_{i=1}^{n+1} a_i \qquad \sigma_B^2 = \sum_{i=1}^{n+1} b_i \qquad \rho \sigma_A \sigma_B = \sum_{i=1}^{n} a_i b_i \tag{17}$$

where σ_A^2 and σ_B^2 are the variances of A and B, respectively, and α denotes the correlation coefficient of A and B.

Then tightness probability (*TA*), which is defined as the probability that arrival time A is larger than B, is computed as:

$$T_A = P(A > B) = \varphi \left(\frac{a_0 - b_0}{\theta} \right) \tag{18}$$

in which:

$$\varphi(y) = \int_{-\infty}^{y} \varnothing(x) \, dx \tag{19}$$

$$\varnothing(x) = \frac{1}{\sqrt{2\pi}} \, e^{-\frac{x^2}{2}} \tag{20}$$

$$\theta = \sqrt{\sigma_A^2 + \sigma_B^2 - 2\rho\sigma_A\sigma_B} \tag{21}$$

Afterward, the mean and variance of $Y = \max(A, B)$ are computed as follows:

$$Y = \max(A, B) = y_0$$
$$+ \sum_{i=1}^{n} y_i \Delta V_{th(i)} + y_{n+1} \Delta V_{th\_rand} \tag{22}$$

$$y_0 = a_0 \, T_A + b_0 \, (1 - T_A) + \theta\varnothing \left(\frac{a_0 - b_0}{\theta} \right) \tag{23}$$

$$\sigma_Y^2 = \left(a_0^2 + \sigma_A^2 \right) T_A + \left(b_0^2 + \sigma_B^2 \right) (1 - T_A)$$
$$+ (a_0 + b_0)\theta\varnothing \left(\frac{a_0 - b_0}{\theta} \right) - y_0^2 \tag{24}$$

Finally, sensitivity coefficients are calculated using the following:

$$y_i = a_i \, T_A + b_i \, (1 - T_A) \tag{25}$$

$$y_{n+1} = \sqrt{\sigma_Y^2 - \sum_{i=1}^{n} y_i^2} \tag{26}$$

2.3   Incremental Criticality-Based Statistical Gate-Sizing Algorithm

In order to guarantee the lifetime constraint of a circuit after several years of aging, considering PV, we propose an incremental criticality-based gate sizing algorithm. Assuming the focus is on meeting constraints on the longest past, it is unnecessary to update the size of a gate on a noncritical path, and because the effect of updating different gates is different and interdependent, it is important to select a group of the most effective gates to be optimized in each iteration of the algorithm. To obtain a candidate group from the most sensitive gates, the gates should be sorted based on the probability that they lie on the critical path(s) due to aging and PV. In order to evaluate how important a gate is for improving reliability, we use the criticality concept [8]. The criticality of a gate is defined as the probability that this gate lies on the critical path due to NBTI and PV.

In a timing optimization flow, after one or more perturbations made to the design, it is only required to update some necessary timing quantities to satisfy a targeted timing parameter. Hence, incremental timing analysis in which only necessary timing quantities are updated instead of all values is a key idea when optimizing and performing a physical synthesis of integrated circuits (ICs). In order to compute the criticality of node e incrementally, all the paths from the source node to the sink node can be divided into two disjoint sets. The first set includes the paths that traverse node e, while the second one includes the paths that are not going across node e. The maximum delay distribution of the paths going through node e is the *node timing slack (NTS)*, which can be expressed as:

$$NTS = AT_e - RAT_e \qquad (27)$$

where AT_e and RAT_e are the arrival time and required arrival time of node e, respectively.

The maximum delay distribution of the paths not traversing node e is called the *complement node timing slack (CNTS)*. Node criticality (c) is defined as the probability of the *NTS* to be greater than or equal to the *CNTS*; that is:

$$c = P(NTS \geq CNTS) \qquad (28)$$

The criticality of a gate can be computed as:

$$\begin{aligned} c = P(NTS \geq CNTS) = P(NTS \geq NTS \ \& \ NTS \geq CNTS) = \\ P(NTS \geq max\,(NTS, CNTS)) = P(NTS \geq C) \end{aligned} \qquad (29)$$

where $C = max\,(NTS, CNTS)$ is the chip timing slack, which is the max delay distribution of all the paths of the circuit from the source node to the sink node or, simply, the AT of the sink node. In other words, the criticality of node e is the probability that its *NTS* dominates the chip timing slack. Thus, we have:

$$c = P(AT_e - RAT_e \geq AT_{sink}) \qquad (30)$$

Therefore, the criticality of gate e, considering NBTI and PV, can be expressed as:

$$c_e(PV, t) = P\{AT_e(PV, \ t) - RAT_e(PV, \ t) \geq AT_{sink}(PV, \ t) \} \qquad (31)$$

where $AT_e(PV, t)$ and $RAT_e(PV, t)$ represent the arrival time and required arrival time of gate e, taking into account NBTI and PV effects, respectively. $AT_{sink}(PV, t)$ represents NBTI- and PV-aware arrival time at the sink node. Therefore, in order to update node criticality c_e, we need to update only AT_e, RAT_e, and AT_{sink}. In order to do this efficiently and incrementally, we only need to run the proposed SSTA for the fan-in cone and fan-out cone of node e. In fact, after updating these quantities incrementally, computing the criticality of node e using Eq. 31 can be done constantly.

Algorithm 1 formulates the proposed gate sizing algorithm. After computing the criticality of each gate in the circuit (lines 4, 5), it ranks the gates according to their criticality (line 6). Then a group of most effective gates is chosen for applying gate-sizing-based optimization according to their optimization priority (lines 7–9). After each iteration of the optimization, SSTA is carried out to obtain the timing of the circuit (lines 13, 14). It is worth noting that during the SSTA procedure, only the fan-in gates of the n-modified gates and those on their fan-out cones need to be reexamined. This iterative sizing procedure terminates when the lifetime reliability constraint is satisfied or the maximal area constraint is violated.

Algorithm 1: Gate Sizing Algorithm
1. Inputs: circuit, lifetime constraint, area constraint
2. Output: optimized circuit
3. n: number of gates chosen for gate sizing
4. **for each** *gate$_i$* **in** a circuit
5. 　　　Compute criticality t_e of *gate$_i$*
6. Rank the gates based on criticality t_e
7. Select the first n gates for gate sizing
8. **for each** *gate$_j$* **in** n
9. 　　　Increase the size of *gate$_j$* by 1
10. Compute the *area* of the optimized circuit
11. **if** the area constraint is achieved
12. 　　　　　**go to** line 18
13. Reexamine the fan-in and fan-out gates of the n-modified gates
14. Compute $(\mu + 3\sigma)$ of the circuit
15. **if** the lifetime constraint is achieved
16. 　　　**go to** line 14
17. **go to** line 4
18. Compute the area overhead of the optimized circuit

3 Experimental Results

The proposed framework is implemented in C++ and is run on a Microsoft Windows machine with an Intel Core i7 (2.4 GHz) processor and 8-GB RAM. Experiments are performed on ISCAS'85 benchmark circuits. Parameters α, A, and γ in the statistical aging model are fitted by HSPICE MOSFET model reliability analysis (MOSRA) simulations under different duty cycles (s), working temperatures (T), and variation distributions of Vth. Initially, all the gates in the circuit are assigned with a base size; e.g., the W/L ratio of PMOS in an inverter is set to 2, and the W/L ratio for its NMOS is set to 1. In addition, for other gates (NAND, NOR, etc.), we have sized the gates such that they have pull-up and pull-down networks with equal current strengths. All of the HSPICE simulations are performed under a PTM 45-nm technology node setting [9].

In order to model the spatial correlations between gates, a multilevel quad-tree partitioning model is used. This is achieved by dividing the die area into several levels and then partitioning each level l into 4^l partitions (i.e., the first or top level, 0, has a single region and the next level, 1, is partitioned into four and so on). Then each partition is assigned an independent random variable. The gates are randomly allocated to a partition at the bottom level, and then the random variable associated with the lowest level partition that contains the gate, along with the random variables associated with the partitions at the higher levels of the quad-tree that intersects the position of the gate, is determined. Here, a quad-tree partitioning model with three levels is used, and all the random variables are set as the standard Gaussian random variable $N \sim (0, 1)$.

3.1 Circuit Lifetime Reliability Optimization

Based on the criticality of the gates, the proposed gate sizing algorithm is performed to optimize the circuit timing. The algorithm terminates when the specified lifetime reliability constraint is satisfied or the maximal area constraint is violated. Here, the lifetime reliability constraint is defined as the $1.1 \times (\mu + 3\sigma)$ of the initial delay of the circuit. In other words, the algorithm is repeated until the $(\mu + 3\sigma)$ of the lifetime circuit delay becomes less than 110% of the initial circuit delay. Also, the area constraint is considered to be 20%. Fifty gates are picked out from the candidate gate group in each iteration according to their criticalities. Area overhead is obtained by calculating the ratio of the total W/L ratios before and after optimization for all the gates in the circuit.

Table 1 illustrates the statistical results after 10 years of lifetime. In this table, the columns named "%mean" and "%variance" respectively represent the mean and variance degradation after 10 years of lifetime, the columns "%mean improve." and "%variance improve." respectively show the percentage of the mean and variance improvement against the lifetime mean and variance, "%area" denotes the

Table 1 Experimental results of gate-sizing-based delay optimization for ISCAS'85 benchmark circuits

Circuit		Before optimization		After optimization			
Bench.	#Gates (#PI,#PO)	%mean	%variance	%mean improve.	%variance improve.	%area	Count
C17	6(5,2)	39.1	−4.88	15.06	43.32	9.55	1
C432	198(36,7)	23.50	−3.1	19.76	35.50	10.25	2
C499	575(41,32)	40.72	−5.32	17.42	26.21	15.75	8
C880	459(60,26)	24.71	−3.15	13.07	27.10	2.8	2
C1355	588(41,32)	39.49	−5.15	10.12	15.63	16.95	6
C1908	524(33,25)	37.84	−4.93	20.96	27.11	16	8
C2670	834(233,140)	31.69	−3.72	17.29	34.42	2.7	2
C3540	1088(50,22)	31.49	−3.93	14.85	27.74	5.063	5
C5315	1666(178,123)	32.07	−3.65	15.25	30.44	2.901	3
C6288	2416(32,32)	39.60	−5.90	13.56	29.54	4.6	4
C7552	3513(207,108)	34.16	−4.36	13.62	27.52	8.5	5
Average		33.99	−4.37	12.41	24.45	8.64	4.18

percentage of area overhead imposed by the sizing procedure, and the column named "count" represents the number of iterations in the sizing procedure. The obtained results show that the proposed optimization technique can achieve the $1.1 \times (\mu + 3\sigma)$ of the initial delay of the circuit constraint with only 8.64% area overhead, on average. It is notable that the amount of lifetime delay mean and variance improvement is larger than the area overhead caused by the optimization algorithm, especially in large circuits. For smaller scale circuits, such as c17, the area overhead is larger due to a smaller number of statistically critical gates with respect to larger circuit delay degradation.

Figure 3 compares the NBTI- and PV-induced lifetime delay mean and variance of each circuit before and after optimization. As can be seen, the optimized circuits achieve timing constraints during their operational lifetime. It is notable that the achieved increase in circuit lifetime reliability is due to the decrease in mean and variance resulting from the gate-sizing-based optimization.

4   Conclusion

In this chapter, we present a statistical framework based on incremental gate sizing to optimize the reliability of a circuit under the joint effects of NBTI and PV. To this end, an aging- and PV-aware canonical first-order delay model is presented to characterize the statistical delay degradation of a gate. Based on the proposed statistical gate delay model, we introduce a block-based SSTA method to estimate the delay of the circuit under the joint effects of PV and NBTI. The proposed

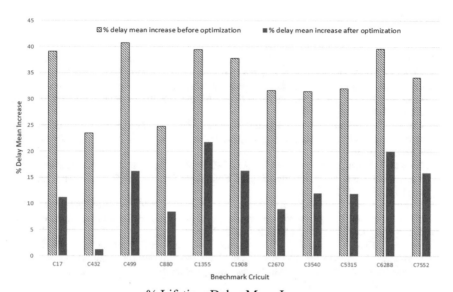

% Lifetime Delay Mean Increase

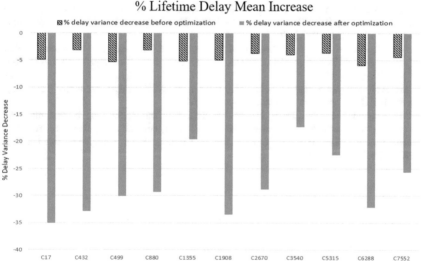

% Lifetime Delay Variance Decrease

Fig. 3 Comparison of lifetime delay mean, variance, and area overhead before and after optimization

statistical timing analysis can be exploited to characterize the lifetime reliability of the circuit with linear runtime complexity. Then a PV- and aging-aware optimization algorithm using an incremental gate sizing approach is proposed to optimize the circuit delay, considering NBTI and PV effects, with low area overhead.

References

1. Oboril F, Aging-aware logic synthesis. In: Proceedings of the International Conference on Computer-Aided Design (ICCAD); 2013. p. 61–8.
2. Jin S, Han Y, Li H, Li X. Statistical lifetime reliability optimization considering joint effect of process variation and aging. Integr VLSI J. 2011;44:185–91. https://doi.org/10.1016/j.vlsi.2011.03.004.
3. Narayan S, Visweswariah C, Ravindran K, Kalafala K, Walker SG. First-order incremental block-based statistical timing analysis. IEEE Trans Comput-Aided Des Integr Circuits Syst. 2006;21:331–6.
4. Sakurai T, Newton AR. Alpha-power law MOSFET model and its applications to CMOS inverter delay and other formulas. IEEE J Solid-State Circuits. 1990;25:584–94. https://doi.org/10.1109/4.52187.
5. Bhardwaj S, Wang W, Vattikonda R, Cao Y, Vrudhula S. Predictive modeling of the NBTI effect for reliable design. In: Proceedings of the custom integrated circuits conference; 2006. p. 189–92. https://doi.org/10.1109/CICC.2006.320885.
6. Kükner H, Khan S, Weckx P, Raghavan P, Hamdioui S, Kaczer B, Catthoor F, Van Der Perre L, Lauwereins R, Groeseneken G. Comparison of reaction-diffusion and atomistic trap-based BTI models for logic gates. IEEE Trans Device Mater Reliab. 2014;14:182–93. https://doi.org/10.1109/TDMR.2013.2267274.
7. Blaauw D, Chopra K, Srivastava A, Scheffer L. Statistical timing analysis: from basic principles to state of the art. IEEE Trans Comput Des Integr Circuits Syst. 2008;27:589–607. https://doi.org/10.1109/TCAD.2007.907047.
8. Xiong J, Zolotov V, Visweswariah C. Incremental criticality and yield gradients. In: Proc Design Autom Test Eur DATE; 2008. p. 1130–5. https://doi.org/10.1109/DATE.2008.4484830.
9. Nanoscale Integration and Modeling (NIMO) Group, ASU, Tempe, Predictive Technology Model (PTM). http://www.eas.asu.edu/~ptm/ (2009)
10. Ebrahimipour SM, Ghavami B, Mousavi H, Raji M, Fang Z, Shannon L. Aadam: a fast, accurate, and versatile aging-aware cell library delay model using feed-forward neural network. In: 2020 IEEE/ACM International Conference on Computer Aided Design (ICCAD); 2020. p. 1–9.

Joint Timing Yield and Lifetime Reliability Optimization of Integrated Circuits

1 Introduction

In order to gain more advances in different aspects, such as performance and power consumption, integrated circuit (IC) manufacturing has followed a trend of transistor dimensions shrinking. As technology continues to scale down, the precision required for fabrication processes has become more difficult to attain, which leads to significant yield loss as well as reduced reliability of ICs. The yield loss of an IC can be divided into catastrophic yield loss, which results from functional failures that cause the part not to work at all, and parametric yield loss, where the chip is functionally correct but fails to meet some power or performance criteria. Parametric failures are caused by variations in one or a set of circuit parameters. Thus, the gap between the designed layout and what is fabricated on silicon is widening significantly. For example, fabrication-process-induced parameter variations (briefly process variations (PVs)) will dramatically affect circuit timing and performance. Process variation (PV) can change circuit frequency by up to 30%. Many ICs are speed-binned, and thus, timing yield, which is one of the most important criteria of parametric yield, has attracted a lot of attention in the literature. The timing yield of the circuit denotes the number of fabricated chips whose delay is satisfied with a specified value called timing constraint. Handling PV effects at design time to improve the timing yield of circuits gives the advantage of making the right decisions early in the design flow, which may significantly reduce the time to market of the ICs. As a result, the impact of parameter variations should be compensated at design time.

Circuit performance not only changes with static process variations that lead to yield issues but also degrades because of runtime variations due to aging effects, resulting in the chip still suffering from reliability problems during its projected lifetime. The negative-bias temperature instability (NBTI) aging phenomenon has been shown to have a significant impact on the lifetime reliability of a circuit. NBTI occurs when a p-channel metal-oxide-semiconductor (PMOS) transistor is negatively biased at an elevated temperature. The degradation process is gradual and

© The Author(s), under exclusive license to Springer Nature Switzerland AG 2023
M. Raji, B. Ghavami, *Lifetime Reliability-aware Design of Integrated Circuits*,
https://doi.org/10.1007/978-3-031-15345-7_5

accumulative over time since not all holes that are trapped can get detrapped, and some of them are lost permanently. Therefore, NBTI manifests itself as an increased threshold voltage shift, which in turn causes the degradation of the device's delay. Since NBTI strongly affects circuit performance during its lifetime, it can cause a violation of timing constraints. To meet the specific timing constraint and increase the reliability of a circuit, it is necessary to consider NBTI-induced delay degradation at design time. Thus, to keep aging effects at bay, designers have to include a timing margin called a timing guardband [1, 2] to ensure the circuit works well for its projected lifetime. Ignoring NBTI-induced degradation results in an incorrect estimation of circuit timing.

The NBTI-induced threshold voltage shift in a transistor depends on both environmental variations (like temperature and input signal probability) and its parameters (like threshold voltage and gate oxide thickness). Since such parameters are variated due to PV effects, the impacts of PV and NBTI have a strong influence on each other. Therefore, the joint effects of NBTI and PV on the circuit design flow should be considered. Among various methods presented to deal with the joint effects of PV and NBTI, gate sizing has shown promising results in nanoscale IC designs.

In this chapter, we introduce a two-phase statistical method using incremental gate sizing to improve both the timing yield and the lifetime reliability of circuits, leading to less area overhead imposed in an optimization algorithm.

2 Problem Formulation

Generally, we have two sources of variation in a circuit: fabrication-induced process variation (PV) and aging-induced variation. Due to PV effects, a circuit may violate the timing constraint, which may lead to a reduction in the timing yield. Therefore, traditional PV-aware optimization techniques aim to optimize the circuit so that delay becomes smaller than the timing constraint. On the other side, when designing a circuit, in order to keep aging effects at bay, designers have to include a timing margin called a timing guard to ensure the circuit works well after a long period of time. Thus, the goal of aging-aware optimization techniques is to optimize the timing guardband.

In other words, traditional aging-aware optimization techniques assume that the initial delay of a circuit is optimized and are focused on optimizing aging-induced variations (timing guardband). As stated in the Introduction, the impacts of PV and aging have a strong influence on each other. Thus, the delay of a circuit at time t, taking into account the joint effects of aging and PV ($D_{circuit}(PV, t)$), can be expressed as:

$$D_{circuit}(PV, t) = D_{circuit}(PV, 0) + \Delta D_{circuit}(PV, t) \tag{1}$$

where, $D_{circuit}(PV, 0)$ is the initial delay of the circuit after chip fabrication ($t = 0$) under PV effects and $\Delta D_{circuit}(PV, t)$ represents aging- and PV-induced lifetime delay degradation.

As a result, in order to keep the circuit working correctly, the circuit delay at time t should be smaller than the timing constraint and the guardband; i.e.:

$$D_{circuit}(PV, 0) + \Delta D_{circuit}(PV, t) \leq T_{clk} + GB \tag{2}$$

where, T_{clk} and GB denote the timing constraint and guardband, respectively.

If the initial delay of the circuit becomes smaller than the timing constraint, the delay degradation of that circuit should be smaller than or equal to the desired guardband in order to have the circuit working correctly in its projected lifetime. We first optimize the initial delay of the circuit such that it becomes smaller than the timing constraint; that is:

$$D_{circuit}(PV, 0) \leq T_{clk} \tag{3}$$

Then the second phase aims to optimize the delay degradation of the circuit obtained from the first phase to become smaller than the guardband (GB); that is:

$$\Delta D_{circuit}(PV, t) \leq GB \tag{4}$$

3   Gate-Level Delay Model Under the Joint Effects of NBTI and PV

In this section, we present a gate-level delay model to characterize process variation and NBTI effects in the delay of a given gate. By extending the statistical gate delay model to the whole circuit, statistical static timing analysis (SSTA) is performed to compute the aging-aware statistical distribution of the circuit delay.

The delay of gate i at time t, considering the joint effects of NBTI and PV ($D_{gate_j}(PV, t)$), can be expressed as follows:

$$D_{gate_j}(PV, t) = D_{gate_j}(PV, 0) + \Delta D_{gate_j}(PV, t) \tag{5}$$

where, $D_{gate_j}(PV, 0)$ is the initial delay of gate j after chip fabrication ($t = 0$) under PV effects and $\Delta D_{gate_j}(PV, t)$ represents aging- and PV-induced lifetime delay degradation.

3.1 Initial Gate Delay Under PV

We will assume that the canonical form of the delay of a gate under PV effects can be stated as a linear function of Gaussian random variables as follows [3]:

$$D_{gate_j} = d_0 + \sum_{i=1}^{n} d_i \Delta V_{th(i)} + d_{n+1} \Delta V_{th\_rand} \tag{6}$$

where, $\Delta V_{th(i)}$ is a normalized random variable representing global V_{th} variations and ΔV_{th_rand} represents the independent variation of the threshold voltage. Meanwhile, d_i and d_{n+1} represent the sensitivity of delay of a gate to each of the random variables ($\Delta V_{th(i)}$ and ΔV_{th_rand}), and d_0 is the mean delay of D_{gate_i}, which can be calculated using the Alpha power law delay model [4]:

$$d_0 = \frac{C_L V_{dd}}{\beta (V_{dd} - V_{th})^\alpha} \tag{7}$$

where, C_L and α respectively represent load capacitance and velocity saturation, V_{th} shows the threshold voltage, β is a parameter that depends on the size of the gate, and V_{dd} represents the supply voltage.

3.2 Delay Degradation Under Joint Effects of NBTI and PV

In order to compute the NBTI- and PV-aware delay degradation of a gate, we first need to compute the impact of NBTI and PV on the V_{th} of the gate. The effect of NBTI on V_{th} can be modeled using the reaction-diffusion model [5] and the trapping and detrapping model [6–8]. A discussion of these modeling theories can be found in Ref. [9]. In this paper, in order to estimate the V_{th} degradation of a PMOS transistor, the widely accepted predictive model of Ref. [5], which is a model based on a reaction-diffusion mechanism is used, i.e. Eq. 8, where K_v is a function of temperature, electrical field, and carrier concentration; β_t is the fraction parameter of the recovery; t represents the transistor lifetime (or the time in which the measurement is performed); V_{th} is the initial threshold voltage of transistor; T_{clk} and s respectively show the input clock cycle and transistor signal probability (duty cycle); and n is determined by dispersion material, which is around 0.16 for H_2:

$$\Delta V_{th\_NBTI}(t) = \left(\frac{\sqrt{K_v^2 \, s \, T_{clk}}}{1 - \beta_t^{\frac{1}{2n}}(t)} \right)^{2n} \tag{8}$$

By incorporating PV-induced uncertainty into the conventional NBTI model [10], we can estimate threshold voltage degradation due to NBTI and PV [11]; that is:

$$\Delta V_{th\_NBTI}(PV, t) = A\left(1 - \gamma\left(\sum_{i=1}^{n} \Delta V_{th(i)} + \Delta V_{th\_rand}\right)\right) s^n t^n \qquad (9)$$

where A is a fitting parameter of V_{th} degradation under nominal conditions. Its value can be extracted from Eq. 8, depending on temperature and supply voltage (V_{dd}). Parameter γ denotes the sensitivity of NBTI-induced V_{th} degradation to the nominal value of V_{th}. In order to compute the delay degradation of a gate taking account of NBTI and PV effects, we transform Eq. 9 into the canonical form as follows:

$$\Delta V_{th\_NBTI} = As^n t^n + \sum_{i=1}^{n} (-As^n t^n \gamma)_i \Delta V_{th(i)} + (-As^n t^n \gamma)_{n+1} \Delta V_{th\_rand} \qquad (10)$$

Delay degradation of a gate is a linear function of threshold voltage shift which can be obtained by computing the derivation from the delay formula considering Alpha power law delay model [4] with respect to the threshold voltage; i.e.:

$$\Delta D_{gate_j}(PV, t) = \mu \Delta V_{th}(PV, t) \qquad (11)$$

where

$$\mu = \alpha D / (V\_(dd) - V\_th0) \qquad (12)$$

Therefore, by substituting ΔV_{th_nbti} (in Eq. 10) with Eq. 11, the delay degradation of a gate, considering the joint effects of NBTI and PV $\left(\Delta D_{gate_j}(PV, t)\right)$, can be expressed as follows:

$$\Delta D_{gate_j}(PV, t) = \mu As^n t^n + \sum_{i=1}^{n} (-\mu As^n t^n \gamma)_i \Delta V_{th(i)} + (-\mu As^n t^n \gamma)_{n+1} \Delta V_{th\_rand} \qquad (13)$$

If D_{gate_j} and ΔD_{gate_j} represent the initial delay and delay degradation of a gate in the canonical form, respectively, we have:

$$D_{gate_j} = d_0 + \sum_{i=1}^{n} d_i \Delta V_{th(i)} + d_{n+1} \Delta V_{th\_rand} \qquad (14)$$

$$\Delta D_{gate_j} = \xi_0 + \sum_{i=1}^{n} \xi_i \Delta V_{th(i)} + \xi_{n+1} \Delta V_{th\_rand} \tag{15}$$

where, ξ_i and ξ_{n+1} represent the sensitivity of delay to each of the random variables and ξ_0 is the mean delay of ΔD_{gate_j}. Thus, the lifetime delay of $gate_j$, considering the joint effects of NBTI and PV, can be expressed as:

$$
\begin{aligned}
D_{gate_j}(PV, t) &= D_{gate_j} + \Delta D_{gate_j} \\
&= x_0 + \sum_{i=1}^{n} x_i \Delta V_{th(i)} + x_{n+1} \Delta V_{t\_rand}
\end{aligned}
\tag{16}
$$

$$x_0 = \xi_0 + d_0 \tag{17}$$

$$x_i = d_i + \xi_i \tag{18}$$

$$x_{n+1} = \xi_{n+1} + d_{n+1} \tag{19}$$

where, x_i and x_{n+1} represent the sensitivity of aging and PV-aware lifetime delay of a gate to each of the random variables, i.e. $\Delta V_{th(i)}$ and ΔV_{th_rand}, and x_0 is the lifetime mean delay of $gate_i$.

4 Gate Sizing Method

In this section, we aim to lessen the impact of aging and process variation using statistical gate sizing. The general idea of the algorithm is to divide the optimization procedure into two phases. In the first phase, using a statistical gate sizing algorithm for improving the timing yield of the circuit, the initial delay of the circuit (i.e., delay of the circuit at $t = 0$) is optimized. Then, in the second phase, we introduce an incremental gate sizing method to guarantee that the circuit meets the timing guardband constraints. In the following, the details of each step have been described. Before explaining the proposed method, some parameters are needed to be introduced:

1. Required arrival time (RAT): a random variable that represents "The time in which data is required to arrive at some internal node of the design."
2. Arrival time (AT): a random variable that represents "The time in which data arrives at the internal node. It incorporates all the net and logic delays in between the reference input point and the destination node."
3. Source node: a virtual node that is connected to all primary inputs of the circuit.
4. Sink node: a virtual node that is connected to all primary outputs of the circuit.

4.1   First Phase: Initial Delay Optimization

To optimize the initial delay of the circuit, we use a statistical gate sizing algorithm for improving the timing yield of the circuit. The timing yield is the number of fabricated chips whose delay is at or below a specified value, called the timing constraint. In other words, the timing yield of a circuit denotes the probability that the circuit slack, which is defined as the difference between the AT and RAT of the circuit or, simply, the AT at the sink node, is nonnegative [12]. This probability can be computed by integrating the circuit slack probability density function (PDF) from 0 to ∞. Given the circuit slack is a Gaussian random variable with mean μ_S and standard deviation σ_S, the timing yield of the circuit (TY), which we attain to improve, is obtained using the following formula:

$$TY = \frac{1}{\sqrt{2\pi}\sigma_S} \int_0^\infty e^{-\frac{(x-\mu_S)^2}{2\sigma_S^2}} dx \equiv \varphi\left(\frac{\mu_S}{\sigma_S}\right) \tag{20}$$

where $\varphi(\cdot)$ represents the standard normal cumulative distribution function (CDF).

It is worthy to note that as the timing yield of the circuit improves, the probability of having a circuit with a delay smaller than or equal to the timing constraint increases. Thus, in order to maximize the metric (μ_S/σ_S), we present a criticality-based gate sizing algorithm. Since it is unnecessary to update the size of a gate on a noncritical path and because the critical path changes from die to die in the presence of PV, we use the criticality concept [13]. The criticality of a given gate is defined as the probability that the gate lies on the critical path taking into account the process variation effects [13].

Without the loss of generality, the incremental criticality computation method introduced in ref [13] is used to find the criticality of each gate. However, any similar approach can be adapted to the proposed algorithm. In a timing optimization flow, after one or more perturbations are made to the design, it is only required to update some necessary timing quantities to satisfy the target timing requirement. Hence, incremental timing analysis, in which only necessary timing quantities are updated instead of all, is a fundament of the optimization and physical synthesis of integrated circuits. In order to compute the criticality of node e incrementally, all of the circuit paths considering the source node and the sink node are divided into two disjoint sets. The first set is covering the paths that traverse node e, while the second one includes the paths that are not going across node e. The maximum delay distribution of the paths going through node e is *Node Timing Slack (NTS)*, which can be expressed as:

$$NTS_e = AT_e - RAT_e \tag{21}$$

where AT_e and RAT_e are the arrival time and required arrival time of node e, respectively.

The statistical maximum delay distribution of the paths that do not traverse node e is called *Complement Node Timing Slack (CNTS)*. Node criticality (c) is defined as the probability that the *NTS* is greater than or equal to the *CNTS*; that is:

$$c = P(NTS \geq CNTS) \tag{22}$$

The criticality of a gate can be computed as [13]:

$$c = P(NTS \geq CNTS) = P(NTS \geq NTS \& NTS \geq CNTS)$$
$$= P(NTS \geq max(NTS, CNTS)) = P(NTS \geq C) \tag{23}$$

where $C = max(NTS, CNTS)$ is the chip timing slack, which is the max delay distribution of all paths of the circuit from the source node to the sink node or, simply, the AT of the sink node. In other words, the criticality of node e is the probability that its NTS dominates the chip slack [13]. Thus, we have:

$$c_e = P(AT_e - RAT_e \geq AT_{sink}) \tag{24}$$

Therefore, the criticality of gate e under process variation effects can be expressed as [13]:

$$c_e(PV, 0) = P\{AT_e(PV, 0) - RAT_e(PV, 0) \geq AT_{sink}(PV, 0)\} \tag{25}$$

In this equation, $AT_e(PV, 0)$ and $RAT_e(PV, 0)$ represent the arrival time and required arrival time of gate e, respectively. $AT_{sink}(PV, 0)$ represents the arrival time at the sink node. Therefore, to update node criticality c, we need to update only AT_e, RAT_e, and AT_{sink}. To do this efficiently and incrementally, we only need to run the proposed SSTA for the fan-in cone and fan-out cone of node e. In fact, after updating these quantities incrementally, computing the criticality of node e using Eq. 25 can be done in constant time.

Algorithm 1 shows the overall gate sizing algorithm. After computing the criticality of each gate in the circuit, the gates are ranked according to their criticality. Since the effect of updating different gates are different and interdependent, it is important to select a group of most effective gates to optimize in each iteration of the algorithm. As a result, a group of eminent gates is chosen for applying the gate sizing procedure. After each iteration of the optimization, SSTA is carried out to obtain the timing of the circuit. It is worthy to note that in the SSTA, only fan-in gates of the n-modified gates and those on their fan-out cones need to be reexamined. This iterated sizing procedure terminates when the timing yield constraint is satisfied or the maximal area constraint is violated.

4.2   Second Phase: Guardband Optimization

After optimizing the initial delay of the circuit, we need to optimize delay degradation so that the lifetime constraint of the circuit is achieved. To this end, we propose a gate sizing algorithm. In the rest of this section, first, we introduce the metrics to provide information for guiding the optimization algorithm. Then we present the ranking method, which is used to rank the gates based on the computed metrics, and finally, the overall algorithm is presented.

4.2.1   Guiding Metrics

During gate sizing, only the gates on critical paths are resized, and it is unnecessary to update the size of a gate on a noncritical path. Since the aging of a gate is dependent on the gate type, temperature, and input signal profile, a path that is labeled noncritical, considering only the impact of PV, might become critical when both PV and aging are considered and vice versa. As a result, it is important to compute the criticality of a gate due to PV and NBTI. We call this metric as aging-aware criticality (ac). It is defined as the probability that a gate lies on a critical path due to NBTI and PV. Therefore, the aging-aware criticality of gate e, considering NBTI and PV, can be expressed as:

$$ac_e(PV, t) = P\{AT_e(PV, t) - RAT_e(PV, t) \geq AT_{sink}(PV, t)\} \qquad (26)$$

where $AT_e(PV, t)$ and $RAT_e(PV, t)$ respectively represent the arrival time and required arrival time of gate e and $AT_{sink}(PV, t)$ represents the arrival time at the sink node.

Since a gate with larger delay degradation has a larger impact on the delay degradation of the paths going through this gate, it is important to optimize the gates with larger delay degradation. Due to PV effects, the delay degradation of each gate is obtained as a random variable represented in canonical form, and, thus, sorting them to constitute the candidate set and provide noteworthy information for guiding statistical optimization is not straightforward.

Algorithm 1: First Phase of Gate Sizing Algorithm
1. *Inputs* : *circuit, timing yield constraint, area constraint*
2. *Output* : *optimized circuit*
3. *n* : *number of gate chosen for gate sizing*
4. **for each** *gate$_i$* **in** *Circuit* **do**
5.   Compute Criticality *t$_i$* of *gate$_i$* (Eq. 25)
6. **end for**
7. **Rank** gates based on *t$_i$*
8. **Select** first *n* gates for gate sizing
9. **for each** *gate$_i$* **in** *n* **do**
10.   **select** the next size of *gate$_i$* from the library cell
11. **end for**
12. Re-examined the fan-in and fan-out gates of *n* modified gates
13. Compute yield $\varnothing\left(\frac{\mu_s}{\sigma_s}\right)$ of the circuit
14. **if** (yield constraint achieved or area constraint violated)
15.   **exit**
16. **else**
17.   **go to line 4**

A prevalent method to transform distribution to value for comparison in sorting purposes is via *projection*. For example, the purpose of conventional *corner projection* is to project each source of variability to one of its two boundary values (generally mean ±3 sigma) and obtain one transformed value for sorting objectives. The *worst-case corner projection*, on the other hand, involves a combination of +3 sigma of each V_{th} source of variability, which jointly pushes the projected delay degradation to the largest possible value. Using the projected value as a surrogate for the comparison may not allow the obtaining of valuable information for steering statistical optimization. Hence, in order to effectively guide statistical design optimization, one could take advantage of statistical approaches rather than worst-case corner projection.

Our proposed sorting approach is based on the concept of tightness probability, i.e., given any two Gaussian random variables A and B, the tightness probability T_A of A is the probability that it is larger than (or dominates) B; i.e. Eq. 27, where μ_A, σ_A. and μ_B, σ_B respectively represent the mean and standard deviation of random variables A and B. $\varphi(\cdot)$ and $\varnothing(\cdot)$ are the standard Gaussian cumulative distribution function (CDF) and standard PDF, respectively. In Eq. 27 below, $\theta = \sqrt{\sigma_A^2 + \sigma_B^2 - 2\rho\sigma_A\sigma_B}$, where $\rho\sigma_A\sigma_B$ is the covariance of A and B:

$$T_A(A, B) = \int_{-\infty}^{\infty} \frac{1}{\sigma_A} \varnothing\left(\frac{X - \mu_A}{\sigma_A}\right) \varphi\left(\frac{\left(\frac{X-\mu_B}{\sigma_B}\right) - \rho\left(\frac{X-\mu_A}{\sigma_A}\right)}{\sqrt{1-\rho^2}}\right) dx = \varphi\left(\frac{\mu_A - \mu_B}{\theta}\right) \quad (27)$$

In order to sort the delay degradation of gate i ($\Delta D_{gate_i}, i = 1...n$), which denotes the random variable computed for circuit gate i, we compute a metric called *aging probability* (ap_i) which is defined as the probability that the delay degradation of the gate is larger than the delay degradation of other gates in the circuit, that is:

$$ap_i = P\left(\Delta D_{gate_i} \geq \max_{j=1...n|j\neq i}\left(\Delta D_{gate_j}\right)\right)$$
$$= T_{\Delta D_{gate}}\left(\Delta D_{gate_i}, \max_{j=1...n|j\neq i}\left(\Delta D_{gate_j}\right)\right) \quad (28)$$

In other words, the aging probability of *gate*$_i$ is the probability that the delay degradation of the gate considering PV effects is the largest among the delay degradation of all other gates under PV effects.

4.2.2   Multiobjective Ranking

In order to optimize the delay degradation of the circuit, we rank the gates based on their aging-aware criticality and aging probability and then choose a group of highest-ranking gates to perform gate sizing. In an ordinary case, we can consider ac_i and ap_i of gate i as two independent probability values. However, since the delay degradation distribution (ΔD) used in aging probability and the distribution function derived from the subtraction of $AT(PV, t)$ and $RAT(PV, t)$ used in aging-aware criticality of a gate share some random variables, they may have a correlation with each other. As a result, using the joint distribution of delay degradation and the distribution function derived from the subtraction of $AT(PV, t)$ and $RAT(PV, t)$, which we call aging aware criticality distribution, for each gate is the best way to describe the behavior of each gate under these distributions.

In order to compute the joint distribution of delay degradation and aging-aware criticality distribution for each gate, we need to compute the correlation between these two distributions. Given $\Delta D_i = \mu_{\Delta D} + \sum_{i=1}^{n} \alpha_i \Delta V_{th(i)} + \alpha_{n+1} \Delta V_{th\_rand}$ and $C_i = \mu_C + \sum_{i=1}^{n} \beta_i \Delta V_{th(i)} + \beta_{n+1} \Delta V_{th\_rand}$ as delay degradation distribution and aging aware criticality distribution of $gate_i$, their 2×2 covariance matrix can be written as:

$$cov(\Delta D, C) = \begin{bmatrix} \alpha_1 & \alpha_2 & \cdots & \alpha_n & \alpha_{n+1} & 0 \\ \beta_1 & \beta_2 & \cdots & \beta_n & 0 & \beta_{n+1} \end{bmatrix} \times V \begin{bmatrix} \alpha_1 & \beta_1 \\ \alpha_2 & \beta_2 \\ \vdots & \vdots \\ \alpha_n & \beta_n \\ \alpha_{n+1} & 0 \\ 0 & \beta_{n+1} \end{bmatrix} \tag{29}$$

where V is the covariance matrix of the sources of variation. Here, we assume that $\Delta V_{th(i)}$ is independent.

As a result, V is the unity matrix, and thus the covariance matrix can be computed using Eq. 30 below, where $\sigma_{\Delta D}^2$ and σ_C^2 respectively represent the variance of delay degradation and aging-aware criticality distribution and $\rho \sigma_{\Delta D} \sigma_C$ also denotes the correlation coefficient of delay degradation and aging-aware criticality distribution:

$$cov(\Delta D, C) = \begin{bmatrix} \sum_{i=1}^{n+1} \alpha_i^2 & \sum_{i=1}^{n} \alpha_i \beta_i \\ \sum_{i=1}^{n} \alpha_i \beta_i & \sum_{i=1}^{n+1} \beta_i^2 \end{bmatrix} = \begin{bmatrix} \sigma_{\Delta D}^2 & \rho \sigma_{\Delta D} \sigma_C \\ \rho \sigma_{\Delta D} \sigma_C & \sigma_C^2 \end{bmatrix} \tag{30}$$

Therefore, we can express the joint effects of delay degradation and aging-aware criticality distributions as a bivariate Gaussian random variable:

$$f_{\Delta D,\ C}(\Delta D, C) \sim N(\mu, \Sigma) \tag{31}$$

where $\mu = \begin{bmatrix} \mu_{\Delta D} \\ \mu_C \end{bmatrix}$ and $\Sigma = cov\ (\Delta D,\ C)$.

After computing the joint distribution of delay degradation and aging-aware criticality for each gate, we need to sort gates based on their joint distribution. So we propose a metric called *delay degradation-aware gate criticality* (DDC_i) for gate i, which is defined as the probability that the joint distribution of delay degradation and aging-aware criticality of gate i is larger than the joint distribution of delay degradation and aging-aware criticality of other gates in the circuit; that is:

$$DDC_i = P\left(f_{\Delta D_i,\, C_i}(\Delta D, C) \geq \max_{\substack{j=1...n \\ j \neq i}} \left(f_{\Delta D_j,\, C_j}(\Delta D, C) \right) \right) \quad (32)$$

However, since computing maximum in Eq. 32 is time-consuming, we compute the probability of the joint distribution of aging and aging aware criticality of gate i being greater than or equal to a reference point for all the gates in the circuit. Since the sink node has maximum delay degradation and aging-aware criticality distribution, we select the sink node as the reference point. So we have:

$$DDC_i = P\left(f_{\Delta D_i,\, C_i}(\Delta D, C) \geq f_{\Delta D_{sink},\, C_{sink}}(\Delta D, C) \right) \quad (33)$$

where $f_{\Delta D_{sink},\, C_{sink}}(\Delta D, C)$ shows the joint PDF of delay degradation and aging-aware criticality distributions. Thus, to compute DDC_i, we need to find the tightness probability for two bivariate Gaussian random variables. Suppose $A \sim N(\mu_A,\ \Sigma_A) = N\left(\mu_{\Delta D_A}, \mu_{C_A}, \sigma^2_{\Delta D_A}, \sigma^2_{C_A}, \rho_A \right)$ be a bivariate Gaussian random variable that represents the joint distribution of aging distribution (ΔD_A) and aging aware criticality distribution (C_A), in which:

$$\Delta D_A = \mu_{\Delta D_A} + \sum_{i=1}^{n} \alpha_{i_A} \Delta V_{th(i)} + \alpha_{n+1_A} \Delta V_{th\_rand} \quad (34)$$

$$C_A = \mu_{C_A} + \sum_{i=1}^{n} \beta_{i_A} \Delta V_{th(i)} + \beta_{n+1_A} \Delta V_{th\_rand} \quad (35)$$

and $B \sim N(\mu_B,\ \Sigma_B) = N\left(\mu_{\Delta D_B}, \mu_{C_B}, \sigma^2_{\Delta D_B}, \sigma^2_{C_B}, \rho_B \right)$ be a bivariate Gaussian random variable that represents the joint distribution of aging distribution (ΔD_B) and aging aware criticality distribution (C_B), in which:

$$\Delta D_B = \mu_{\Delta D_B} + \sum_{i=1}^{n} \alpha_{i_B} \Delta V_{th(i)} + \alpha_{n+1_B} \Delta V_{th\_rand} \quad (36)$$

$$C_B = \mu_{C_B} + \sum_{i=1}^{n} \beta_{iB} \Delta V_{th(i)} + \beta_{n+1_B} \Delta V_{th\_rand} \tag{37}$$

Given any two bivariate Gaussian random variables $A \sim N(\mu_A, \Sigma_A)$ and $B \sim N(\mu_B, \Sigma_B)$, the tightness probability T_A of A is the probability that it is larger than (or dominates) B is:

$$DDC_i = P(A \geq B) = P(A - B \geq 0) \tag{38}$$

As a result, we need to compute the subtraction of A from B which is, in turn, a bivariate Gaussian random variable, that is:

$$(A - B) \sim N(\mu_{SUB}, \Sigma_{SUB}) = N\left(\mu_{\Delta D_{SUB}}, \mu_{C_{SUB}}, \sigma^2_{\Delta D_{SUB}}, \sigma^2_{C_{SUB}}, \rho_{SUB}\right) \tag{39}$$

$$\mu_{\Delta D_{SUB}} = \mu_{\Delta D_A} - \mu_{\Delta D_B} \tag{40}$$

$$\mu_{C_{SUB}} = \mu_{C_A} - \mu_{C_B} \tag{41}$$

$$\sigma^2_{\Delta D_{SUB}} = \sigma^2_{\Delta D_A} + \sigma^2_{\Delta D_B} - 2\rho_{\Delta D_A, \Delta D_B} \sigma_{\Delta D_A} \sigma_{\Delta D_B}$$
$$= \sum_{i=1}^{n+1} \alpha^2_{i_A} + \sum_{i=1}^{n+1} \alpha^2_{i_B} - 2 \sum_{i=1}^{n} \alpha_{i_A} \alpha_{i_B} \tag{42}$$

$$\sigma^2_{C_{SUB}} = \sigma^2_{C_A} + \sigma^2_{C_B} - 2\rho_{C_A, C_B} \sigma_{C_A} \sigma_{C_B}$$
$$= \sum_{i=1}^{n+1} \beta^2_{i_A} + \sum_{i=1}^{n+1} \beta^2_{i_B} - 2 \sum_{i=1}^{n} \beta_{i_A} \beta_{i_B} \tag{43}$$

$$\rho_{SUB} = \left(\sum_{i=1}^{n} (\alpha_{i_A} - \alpha_{i_B})(\beta_{i_A} - \beta_{i_B})\right) / (\sigma_{\Delta D_{SUB}} \sigma_{C_{SUB}}) \tag{44}$$

Therefore, DDC_i can be expressed as:

$$DDC_i = P(A \geq B) = P(A - B \geq 0)$$
$$= 1 - P(A - B < 0) = 1 - F_{\Delta D_{SUB}, C_{SUB}}(0, 0) \tag{45}$$

After computing DDC_i of each gate i using Eq. 45, the gates are ranked according to their DDC_i. Then a group of gates with the highest level of DDC is chosen for the application of gate-sizing-based optimization.

Algorithm 2 shows the overall gate sizing algorithm. After computing the DDC of each gate in the circuit, the gates are ranked according to their DDC. Since the effect of updating different gates is different and interdependent, it is important to select a group of most effective gates to be optimized in each iteration of the algorithm. As a result, a group of eminent gates is chosen for applying the gate sizing procedure. After each iteration of the optimization, SSTA is carried out to obtain lifetime delay of the circuit. It is noteworthy that in the SSTA, only the fan-in gates of n-modified gates and those on their fan-out cones need to be reexamined. This iterated sizing procedure terminates when the lifetime reliability constraint is satisfied or the maximal area constraint is violated.

Algorithm 2: Second Phase of Gate Sizing Algorithm
  1. ***Inputs*** : *circuit, lifetime reliability constraint, area constraint*
  2. ***Output*** : *optimized circuit*
  3. *n* : *number of gate chosen for gate sizing*
  4. **for each** *gate$_i$* **in** *Circuit* **do**
  5.   Compute Delay Degradation-aware gate Criticality
                 DDC$_i$ of *gate$_i$* (Eq. 45)
  6. **end for**
  7. **Rank** gates based on *DDC$_i$*
  8. **Select** first *n* gates for gate sizing
  9. **for each** *gate$_i$* **in** *n* **do**
  10.   **select** the next size of *gate$_i$* from library cell
  11. **end for**
  12. Reexamined the fan-in and fan-out gates of *n* modified gates
  13. Compute lifetime delay of the circuit
  14. **if** (lifetime reliability constraint achieved or area constraint violated)
  15.   **exit**
  16. **else**
  17.   **go to line 4**

5   Experimental Results

The proposed framework is implemented in C++ and is run on a machine with an Intel Core i7 (2.5 GHz) processor and 8-GB RAM. Experiments are performed on ISCAS'85 and EPFL benchmark circuits. Initially, all the gates in the circuit are assigned with a base size; e.g., the W/L ratio of PMOS in an inverter is set at 2, and the W/L ratio for its NMOS is set at 1. In addition, for other gates (NAND, NOR, etc.), we have sized them in a way that they have pull-up and pull-down networks with equal current strengths.

To model the spatial correlations between gates, a multilevel quad-tree partitioning model is used [14]. This is achieved by dividing the die area into several levels and then partitioning each level l into 4^l partitions (i.e., the first or top level, 0, has a single region, and the next level, 1, is partitioned into 4 and so on). Then each partition is assigned an independent random variable. The gates are allocated to a partition in the bottom level and then, the random variable associated with the lowest level partition that contains the gate along with the random variables associated with the partitions at the higher levels of the quad-tree that intersects the position of the gate are determined. Here, we assume variations on threshold voltage, and a quad-tree partitioning model with three levels is used. All random variables are set to be the standard Gaussian random variable with a mean (μ) of -0.22(V), and the 3σ point for each process parameter is 10% of its mean [15], i.e., a variation ratio ($3\sigma/\mu$) of 10%.

5.1 Effect of Timing Yield Optimization

In order to investigate the efficiency of the timing yield optimization phase of the proposed gate sizing algorithm, an experiment is conducted in which we compare the area overhead imposed by the proposed two-phase optimization algorithm (Table 1) with the state-of-the-art algorithm, which only optimizes the lifetime delay of the circuit, taking into account aging and PV effects based on the DDC metrics. The lifetime reliability constraint of the state-of-the-art algorithm is defined as the one used in Table 1. In each iteration of the algorithm, the gates with DDC greater than 0.97 of the maximum DDC are chosen for optimization.

5.2 Evaluation of the Delay Degradation-Aware Gate Criticality Metric

Table 2 compares the area overhead, runtime, and number of iterations needed for the convergence of the optimization algorithm using the gate sizing method based on the proposed metric (i.e., DDC) and the aging-aware gate criticality.

The gate sizing algorithm is applied until the lifetime delay of the circuit becomes equal to the lifetime reliability constraint, which is defined as 120% of the initial circuit delay; i.e., the design guardband is set to 20%. In each iteration, a group of the gates is picked out from the candidate gate set, according to their optimizing priorities. It is notable that the group size is a parameter that provides the designer with a tradeoff between the optimization runtime and its accuracy (i.e. by choosing smaller group size, the final solutions would be better in the expense of more runtime). The effectiveness of each metric is assessed based on the imposed area overhead needed to achieve the specified lifetime reliability constraint. As reported in this table, on average, the area overhead imposed by the DDC-based gate sizing is 13.45%, which is lower than the 39.03% area overhead imposed by the aging-aware criticality-based gate sizing. The comparison demonstrates that DDC is an effective metric for guiding aging-aware statistical lifetime reliability optimization as it leads to lower area overhead in comparison with the conventional aging-aware criticality metric.

Finally, statistical gate sizing is performed on the group of the highest ranking gates such that the circuit satisfies the lifetime reliability constraint. For all methods, a dynamic threshold that is a fraction ($k = 0.97$) of the maximum of the guiding

Table 1 Experimental results of gate-sizing-based delay optimization for ISCAS'85 and EPFL benchmark circuits.

Benchmark circuit		Before optimization		Timing yield optimization		Lifetime reliability optimization		Total		
Bench	#Gates (#PI, # PO)	$\%\mu$	$\%\sigma^2$	%area	count	%area	count	%area	count	%power
C432	198(36,7)	23.5	−3.1	2.14	1	2.18	1	4.32	2	3.86
C499	575(41,32)	35.7	−5.32	3.38	1	6.84	2	10.22	3	12.94
C880	459(60,26)	24.7	−3.15	3.92	1	3.97	1	7.89	2	6.17
C1355	588(41,32)	34.4	−5.15	7.12	2	18.61	5	25.73	7	21.42
C1908	524(33,25)	33.8	−4.93	5.19	1	9.44	2	14.63	3	11.95
C2670	834(233,140)	31.6	−3.72	3.61	1	3.87	1	7.48	2	7.21
C3540	1088(50,22)	31.4	−3.93	5.24	2	5.68	2	10.92	4	8.46
C5315	1666(178,123)	32.0	−3.65	2.52	1	2.98	1	5.5	2	4.68
C6288	2416(32,32)	35.6	−5.90	17.6	4	10.98	3	28.58	7	17.53
C7552	3513(207,108)	34.1	−4.36	9.24	2	16.92	3	26.16	5	16.59
Max	5824(512,130)	33.5	−4.43	3.58	1	3.23	1	6.81	2	8.71
Sin	11031(24,25)	34.3	−4.7	5.81	1	8.84	2	14.65	3	14.31
Arbiter	23747(256,129)	36.7	−5.57	6.38	2	9.33	3	15.71	5	14.16
Voter	31622(1001, 1)	35.1	−5.13	4.46	1	3.92	1	8.38	2	5.52
Square	39558(64,128)	36.2	−5.4	2.72	1	3.12	1	5.84	2	5.37
Sqrt	54215(128, 64)	34.8	−4.94	6.28	2	13.68	4	19.96	6	16.34
Multiplier	55910(128, 128)	35.5	−5.31	8.58	2	7.76	2	16.34	4	12.47
Average		33.1	−4.37	5.86		7.84		13.72		11.04

Table 2 Comparison of area overhead, runtime, and number of iterations needed for the convergence of the optimization algorithm based on the delay degradation-aware gate criticality and aging-aware criticality metrics

Benchmark circuit	%area		Iterations		Runtime(s)	
	DDC	Criticality	DDC	Criticality	DDC	Criticality
C432	2.23	4.34	1	2	0.1	0.2
C499	6.94	13.8	2	4	0.3	6.04
C880	8.02	12.15	2	3	1.04	1.36
C1355	52.7	101.84	10	19	8.4	15.96
C1908	18.44	52.58	4	11	4.2	11.55
C2670	15.28	23.46	4	6	6.2	9.3
C3540	8.13	23.85	3	9	5.37	16.11
C5315	6.04	20.23	2	7	7	24.5
C6288	14.84	41.25	4	11	14.92	41.03
C7552	28.65	78.26	5	14	26.4	73.92
Max	6.68	16.95	2	5	21.4	53.5
Sin	8.76	67.65	2	15	48.56	364.2
Arbiter	12.08	37.44	4	12	506.68	1520.04
Voter	8.06	73.26	2	18	597	5373
Square	3.05	25.2	1	8	403.51	3228.08
Sqrt	17.6	36.1	5	10	4282.8	8565.6
Multiplier	11.07	35.19	3	9	2678.67	8036.01
Average	**13.45**	**39.03**			**565.44**	**1608.26**

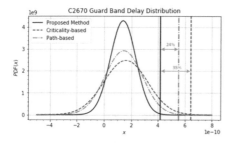
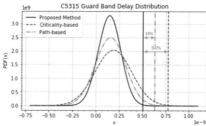

Fig. 1 Comparison of the timing guardband distribution imposed by the proposed gate sizing method and the optimization method based on the criticality-based and path-based methods for C2670 and C5315 benchmark circuits

metric, i.e. criticality and worst-case delay degradation, is used to pick out the gates from the candidate gate group in each iteration given their optimizing priorities. The algorithms stop when they achieve either the lifetime reliability constraint, which is 110% of the initial delay distribution of the circuit, or the runtime constraint, which is 10 h (i.e., the algorithm will terminate when its runtime exceeds 10 h). Since the results of the proposed method itself have entries close to 30%, we consider providing results at a higher area increase threshold (i.e., 40% area overhead) as it shows how much better these values are in comparison to the state-of-the-art methods. Figure 1 compares the timing guardband delay distribution of the circuit (C2670 and C5315 as some examples) after optimizing with the proposed method and criticality-based and path-based ones. As can be seen, for C2670 benchmark circuit, the proposed method leads to 24% and 35% smaller timing guardband compared to the path-based and criticality-based optimization approaches, respectively. The timing guardband imposed by the proposed method for the C5315 benchmark circuit is 19% and 35% smaller than the timing guardband imposed by the path-based and criticality-based methods, respectively.

6   Conclusion

In this chapter, a two-phase statistical method using incremental gate sizing was introduced to address the postfabrication timing yield and lifetime reliability issues of circuits. In the first phase, timing yield is improved by optimizing the initial delay of the circuit. To this end, we first compute the criticality of each gate, and then the gates are ranked based on their criticalities. Finally, in order to have an efficient gate sizing, a group of gates with the highest level of criticality is chosen for gate-sizing-based optimization. In the second phase of the proposed method, in order to guarantee that the circuit meets the lifetime reliability constraint, we introduce an incremental gate sizing method in which a novel concept called aging probability is introduced. Using aging probability and aging-aware criticality, we propose a novel concept called delay degradation-aware gate criticality, which takes advantages of an adaptive multiobjective ranking approach.

References

1. Amrouch H, Khaleghi B, Gerstlauer A, Henkel J. Reliability-aware design to suppress aging. In: Proceeding of the Design Automation Conference (DAC); 2016. p. 1–6.
2. Ebrahimipour SM, Ghavami B, Raji M. A statistical gate sizing method for timing yield and lifetime reliability optimization of integrated circuits. IEEE Trans Emerg Topics Comput. 2021;9(2):759–73. https://doi.org/10.1109/TETC.2020.2987946.
3. Narayan S, Visweswariah C, Ravindran K, Kalafala K, Walker SG. First-order incremental block-based statistical timing analysis. IEEE Trans Comput-Aided Design Integr Circuits Syst. 2006;25(10):331–6.
4. Sakurai T, Newton AR. Alpha-power law MOSFET model and its applications to CMOS inverter delay and other formulas. IEEE J Solid-State Circuits. 1990;25:584–94.
5. Bhardwaj S, Wang W, Vattikonda R, Cao Y, Vrudhula S. Predictive modeling of the NBTI effect for reliable design. In: Proc IEEE Custom Integr Circuits Conf (CICC); 2006. p. 189–92.
6. Grasser T, Member S, Kaczer B, Goes W, Reisinger H, Aichinger T, Hehenberger P, Wagner P, Schanovsky F, Franco J, Luque MT, Nelhiebel M. The paradigm shift in understanding the bias temperature instability : from reaction – diffusion to switching oxide traps. IEEE Trans Device Mater Rel. 2011;58(11):3652–66.
7. Kaczer B, Franco J, Weckx P, Roussel PJ, Putcha V, Bury E, Simicic M, Chasin A, Linten D, Parvais B, Catthoor F, Rzepa G, Waltl M, Grasser T. A brief overview of gate oxide defect

properties and their relation to MOSFET instabilities and device and circuit time-dependent variability. Microelectron Reliab. 2018;81:186–94.

8. Both TH, Firpo Furtado G, Wirth GI. Modeling and simulation of the charge trapping component of BTI and RTS, Microelectron. Microelectron Reliab. 2018;80:278–83.

9. Kükner H, Khan S, Weckx P, Raghavan P, Hamdioui S, Kaczer B, Catthoor F, Van Der Perre L, Lauwereins R, Groeseneken G. Comparison of reaction-diffusion and atomistic trap-based BTI models for logic gates. IEEE Trans Device Mater Rel. 2014;14(1):182–93.

10. Agarwal A, Blaauw D. Statistical timing analysis for intra-die process variations with spatial correlations. In: Proc IEEE/ACM Int Conf Comput Aided Design (ICCAD); 2003. p. 900–7.

11. Jin S, Han Y, Li H, Li X. Statistical lifetime reliability optimization considering joint effect of process variation and aging. Integr VLSI J. 2011;26(4):185–91.

12. Sinha D, Shenoy NV, Zhou H. Statistical timing yield optimization by gate sizing. IEEE Trans Very Large Scale Integr (VLSI) Syst. 2006;14(10):1140–6.

13. Xiong J, Zolotov V, Visweswariah C. Incremental criticality and yield gradients. In: Proceedings of the conference on design, automation and test in Europe (DATE); 2008. p. 1130–5.

14. Agawal A, Blaauw D, Zolotov V. Statistical timing analysis for intra-die process variations with spatial correlations. In: Proc IEEE/ACM Int Conf Comput Aided Design (ICCAD); 2003. p. 900.

15. Ramprasath S, Vasudevan V. Efficient algorithms for discrete gate sizing and threshold voltage assignment based on an accurate analytical statistical yield gradient. ACM Trans Design Autom Electron Syst (TODAES). 2016;21(4):Article 66. 27 pages

Lifetime Reliability Optimization Algorithms of Integrated Circuits Using Dual-Threshold Voltage Assignment

1 Introduction

Although technology scaling allows us to fabricate chips with higher complexity and performance, it poses a severe challenge in creating reliable digital circuit designs. With the scaling down of transistor dimensions, fabrication-induced process variation (PV) causes the timing of the manufactured circuit to significantly deviate from its initial design [1, 2]. As an example, with process technology scaling from 350 nm to 45 nm, the timing yields of the integrated circuits (ICs) have decreased from nearly 90% to a mere 30% [3]. On the other hand, the aging mechanisms, such as bias temperature instability (BTI), are appeared in nanoscale technology leading to considerable degradation of the conductance of metal-oxide-semiconductor (MOS) transistors during lifetime [2, 4]. Negative BTI occurring in P-channel metal-oxide-semiconductor (PMOS) transistors and positive BTI affecting N-channel metal-oxide-semiconductor (NMOS) transistors increase the absolute value of the threshold voltage (Vth) of transistors. Hence, circuit delay increases with operation time. Consequently, it is necessary to consider the effects of PV and aging (specifically BTI mechanisms) in analyzing and improving the lifetime reliability of nanoscale digital circuits.

Considering the interdependencies between BTI and PV [2, 5], a separate analysis of PV and aging leads to inaccurate and unrealistic results. Recently, some papers have addressed the combined impacts of process variability and negative-bias temperature instability (NBTI) or BTI at different levels of abstraction, including at the gate level. For example, in Ref. [6], a statistical circuit optimization flow was introduced, which improves the lifetime reliability of combinational circuits in the presence of the joint effects of PV and NBTI. Khan et al. [7] presents a simulation-based (HSPICE Monte Carlo) analysis of both NBTI and PBTI incorporate with parameter variations in logic gates. Lu et al. [8] have analyzed the impact of PV and NBTI effects on the lifetime reliability of ICs. Although the aforementioned works considered the joint effects of NBTI and PV, their analysis is limited to NBTI and does not cover both NBTI and PBTI, while it has been shown that PBTI is changing to a severe aging mechanism in modern very-large-scale integrated (VLSI) systems [9]. In Ref. [10], a new two-phase gate-sizing approach is presented to improve the reliability of circuits, taking into account the joint effects of process variation and transistor aging. In the first stage, the initial delay of the circuit is optimized to improve its timing yield, and then in the second stage, delay degradation induced by aging and process variations is reduced using gate sizing. Some researchers have studied the effects of both NBTI and PBTI on circuit reliability. In Ref. [9], an analytical methodology for the accurate modeling of the correlation between process, voltage, and temperature (PVT) variations and BTI-induced aging is presented. However, most of the aforementioned works use the old reaction-diffusion theory to consider NBTI and PBTI effects on circuits.

There are several approaches presented for mitigating the impact of either PV [11, 12] or aging [13, 14] in combinational circuits. Due to the interdependencies between PV and aging [15, 16], the mitigation technique, which ignores the PV or aging effects, results in nonoptimized solutions. Among the mitigation approaches in which both PV and aging effects are taken into account, in the literature, gate-sizing-based optimization techniques show their efficacy. In Ref. [7], mathematical methods (i.e., Lagrangian relaxation) are used to optimize the circuit area, taking into account circuit delay degradation. However, the computational complexity of these methods makes them impracticable for large-scale combinational circuits. Other approaches are based on upsizing a gate for delay degradation reduction in the expense of area cost [6, 17]. However, gate upsizing is the main idea of the aforementioned works for improving the delay in the critical path, which results in a considerable area cost. Gomez et al. [13] propose a method for reducing guardband through an effective selection of critical gates and by using a gate-sizing approach that considers the BTI and PV effects. This work focuses on finding a potential critical path (PCP) (paths which are degraded more than other ones) to up/downsizing gate in the path in terms of low area overhead. However, since this work only uses gate sizing technique, it leads to a large area overhead to the VLSI designs.

In this chapter, we present an optimization framework for maximizing circuit lifetime reliability in the presence of PV and aging effects using the dual-threshold voltage (DVth) technique [18]. Firstly, we introduce a novel lifetime reliability evaluation metric to determine the reliability of combinational circuits. Then a statistical gate delay model is presented to predict circuit reliability at the end of its lifetime, considering the PV, NBTI, and PBTI effects. We present a Guardband-Aware Reliability (GAR) metric to analyze the lifetime reliability of combinational circuits using a guardband and timing yield specified by the designer. Furthermore, we introduce a criticality metric to efficiently identify the best candidate gates for applying the DVth assignment technique. Then in order to improve the lifetime reliability of combinational circuits in the presence of PV and aging effects, three optimization algorithms are presented:

- Greedy-based reliability optimization algorithm (abbreviated as GeRO)
- Metaheuristic-based reliability optimization approach using the simulated-annealing (SA) technique (abbreviated as SARO)
- Sensitivity-based reliability improvement algorithm based on the TILOS approach [19] (abbreviated as TIRO)

In these three optimization algorithms, there is a tradeoff between optimization improvement and optimization speed. GeRO and SARO are fast, while TIRO achieves more reliability improvement due to the presented sensitivity metric. On average, the experimental results on International Symposium on Circuits and Systems (ISCAS)-85 and ISCAS-89 benchmark circuits show that the presented method increases circuit reliability by up to 6.38%, 8.16%, and 9.93%, imposing 6.50%, 7.03%, and 6.9% timing yield loss for a 6-year lifetime and 10% PV for GeRO, SARO, and TIRO, respectively.

2 PV- and BTI-Aware Gate Delay Model

In this chapter, a gate delay model is presented in order to consider the impact of PBTI [18]. The PV- and NBTI/PBTI-aware delay of gate k (D_k) is considered as:

$$D_k = D_{nom(k)} + B_k.a_k^n.t^n + \left(1 - \gamma.A.a_k^n.t^n\right).\beta_k.\sum_i \Delta V_{th(i)} \qquad (1)$$

where $D_{nom(k)}$ is the nominal gate delay and B_k is a fitting parameter, which shows increasing gate delay caused by a BTI-induced threshold voltage increase under nominal conditions. Meanwhile, β_k is a fitted coefficient, which indicates the PV-induced threshold voltage variation effects on the gate delay variation without considering the BTI effect.

A BTI-induced Vth increase depends on the fraction of time the transistor is under stress (i.e., an input signal is logic "0" for PMOS and "1" for NMOS) for a period of time, and it is called stress probability (SP). Note the difference between SP and statistical signal probability, which is typically defined as the probability that the input signal becomes logic "1."

Due to the complementary behavior of the pull-up and pull-down network of CMOS gates, the gate is under maximum stress when the signal probability. To consider the impact of both NBTI and PBTI, the a value is computed as:

$$a = |0.5 - logic\ gate\ inputs\ signal\ probablity| \qquad (2)$$

Note that Vth contains two deviation parts; the first part is associated with the time-zero variation, and the second part is associated with aging effects (see Eq. 1). The computational complexity of this model is low, and the error spread by discarding high-order terms in this linear model can be ignored.

Before doing statistical static timing analysis (SSTA), the parameters in Eq. 2 are precomputed by using HSPICE simulations for basic gate types (i.e., INV, BUFF, NAND, NOR) at different designs.

3　Guardband-Aware Lifetime Reliability (GAR) Metric

In this section, we present a novel metric, called Guardband-Aware timing Reliability (GAR), to analyze and evaluate the lifetime reliability of combinational circuits.

In order to obtain the GAR metric, we first define the concept of lifetime reliability for a combinational circuit. Based on the traditional formal definition of reliability, combinational circuit lifetime reliability at time t is defined as the probability of operating circuit at time t correctly given the circuit works correctly at time 0 (fresh time)); i.e.:

$$\mathcal{R}(t) = \mathcal{P}(operational\ at\ t|operational\ at\ 0) \tag{3}$$

A combinational circuit works properly at time 0 when the delay of the critical path (CP) is less than a given timing constraint (τ). A circuit is reliable at the end of a specific lifetime if the CP delay at that time (t) remains less than the same timing constraint at the fresh time (i.e., τ). Hence, Eq. 3 can be rewritten as:

$$\mathcal{R}(t) = \mathcal{P}(d_t < \tau|\ d_0 < \tau) \tag{4}$$

where d_t and d_0 respectively represent the values of circuit CP delay at time t and time 0, and τ is the timing constraint of the combinational circuit defined by the designer at the design stage.

In order to solve Eq. 4, we use the conditional probability calculation rules. Hence, we have:

$$\mathcal{R}(t) = \frac{\mathcal{P}(d_t < \tau \cap d_0 < \tau)}{\mathcal{P}(d_0 < \tau)} \tag{5}$$

where $\mathcal{P}(d_t < \tau \cap d_0 < \tau)$ shows the probability of the intersection of two events: $d_t < \tau$ and $d_0 < \tau$ (i.e., combinational circuit CP delay at time τ and time 0 being less than the timing constraint τ). Since CP delay increases by increasing the operation time of the combinational circuit due to the BTI effects, we have:

$$d_0 < d_t \tag{6}$$

Therefore, if event $d_t < \tau$ occurs, event $d_0 < \tau$ has certainly occurred. So we have:

$$d_0 < \tau \subset d_t < \tau \tag{7}$$

where $\subset$ shows that event $d_0 < \tau$ is a subset of event $d_t < \tau$. Based on the rules of probability calculation, we have:

$$P(d_t < \tau \cap d_0 < \tau) = P(d_t < \tau) \tag{8}$$

Substituting Eq. 8 with Eq. 5, lifetime reliability is obtained as follows:

$$R(t) = \frac{P(d_t < \tau)}{P(d_0 < \tau)} \tag{9}$$

Due to the impacts of variabilities, combinational circuit CP delay is modeled as a random variable with Gaussian distribution [20]. So $P(d_0 < \tau)$ is equal to the value of the cumulative distribution function (CDF) of CP delay at time 0 for timing constraint τ. Designers specify the variation-aware timing constraint for combinational circuits by using the concept of p-percentile point of CP delay CDF, i.e., the value for which the CDF is equal to p.

Designers of high-reliable circuits consider a guardband for the obtained p-percentile point value of CP CDF to ensure they obtain the targeted lifetime reliability [14]. So for reliability-aware circuit designs, timing constraint τ is computed as:

$$\tau = (1 + g) \times \varphi_0^{-1}(p/100) \tag{10}$$

where φ_0^{-1} shows the inverse CDF of CP delay at time 0 and g shows the guardband value ($0 \le g \le 1$) considered by the designer. Therefore, Eq. 9 can be reexpressed as:

$$R_g^p(t) = \frac{\varphi_t\big((1 + g) \times \varphi_0^{-1}(p/100)\big)}{\varphi_0\big((1 + g) \times \varphi_0^{-1}(p/100)\big)} \tag{11}$$

where $R_g^p(t)$ is the GAR metric for the lifetime reliability of the combinational circuit at time t considering the guardband value of g for the p-percentile point value of CP CDF, φ_t and φ_0 show the CDF of CP delay at time t and 0, respectively.

The GAR degradation for a specific guardband g and p-percentile point value at operational time t ($\Delta R_g^p(t)$) is computed as:

$$\Delta R_g^p(t) = \frac{R_g^p(t) - R_g^p(0)}{R_g^p(0)} \tag{12}$$

4　Dual-Threshold Voltage Assignment Technique

Dual-threshold voltage assignment (DVth) is widely used for low-voltage, low-power, and low-leakage power applications [21]. In dual Vth designs, higher Vth is assigned to the gates in noncritical paths for reducing leakage power, and lower Vth is assigned to the gates in critical paths for improving performance/timing yield. Implementing the dual Vth technique is easy to fabricate using an additional mask layer [21]. Here, we present applying the dual Vth assignment technique for reducing the impact of BTI on circuit delay for improving lifetime reliability. A transistor with higher Vth is expected to experience less BTI effect (Vth shift) due to the reduction in the electric field stress in the oxide (*Eox*) consign [21]. Here we provide a motivation example to show the DVth effect on BTI-induced circuit delay degradation.

4.1　Motivation Example

Figure 1 shows the C17 circuit of ISCAS'85 benchmark suite. We conducted a set of experiments in which three different Vths are assigned to all the gates in the circuit, and each time, we measure the delay of the highlighted path (from input node N1 to output node N22). Figure 2 shows the GAR values ($\mathcal{R}_{-0.1}^{0.99}(t)$ for $0 \leq t \leq 10$) of the C17 circuit with three different Vths. It is observed that higher Vth leads to higher lifetime reliability. For example, the reliability of the circuit after its 10-year lifetime is around 0.86 when Vth of 660 mV is assigned to the gates, while it is reduced to 0.8 for Vth equal to 480 mV.

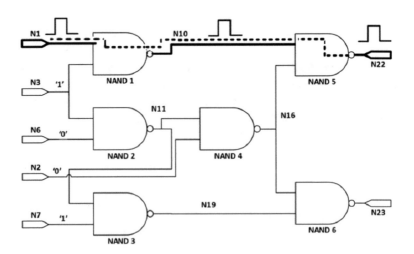

Fig. 1　C17 benchmark structure

Fig. 2 GAR value of the highlighted path in Fig. 1 for different Vth values during a 10-year lifetime

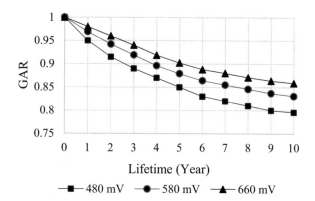

Fig. 3 Delay of the highlighted path in Fig. 1 for different Vths

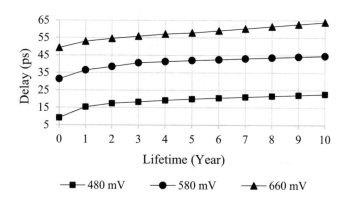

4.2 Vth Assignment Technique Overhead

Although higher Vth results in less BTI-induced reliability degradation, using higher Vth imposes delay overhead on the design, leading to timing yield loss in the circuit. Figure 3 shows the delay of the path considered in the motivation example during the 10 years of circuit lifetime. It is observed that the delay of the circuit is increased with the increase of the lifetime.

In order to formally evaluate the cost of the DVth assignment technique, we describe the cost function ($Cost_g^p$) as follows:

$$Cost_g^p = \frac{\varphi_L\big((1+g) \times \varphi_L^{-1}(p/100)\big) - \varphi_H\big((1+g) \times \varphi_L^{-1}(p/100)\big)}{\varphi_L\big((1+g) \times \varphi_L^{-1}(p/100)\big)} \tag{13}$$

Table 1 Cost and GAR degradation for different gate selections

DVth gate	Initial delay (ps)	Cost (%)	$\Delta \mathcal{R}_{0.1}^{99}(10)$ (%)
No gate	9.13	0	49
All gates in the path	31.38	73	11
NAND 5	17.45	8	14

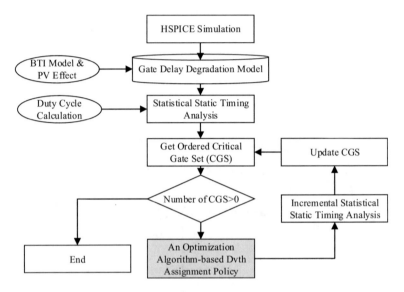

Fig. 4 Overall flow of the presented reliability optimization framework

where φ_L and φ_H show the CDF of CP delay at time 0 for low and high Vth designs, respectively, and φ_L^{-1} indicates the inverse CDF of CP delay at time 0 for low Vth circuits. The cost function reflects the timing yield loss due to increased delay caused by the DVth technique.

In order to investigate the overhead of the DVth assignment technique, we provide an example in which DVth is applied to C17 in different cases. The obtained results of these experiments are presented in Table 1. The first column shows the different cases of applying the DVth assignment technique; i.e., "No Gate" refers to a case in which no gates are assigned with high Vth, "All gates in the path" indicates a case in which all the gates in the highlighted path in Fig. 4 are assigned with high Vth and the other ones have low Vth, and "NAND 5" shows a case in which only gate NAND 5 is assigned with high Vth and the other ones have low Vth. The other columns show the initial delay of the circuit (99-percentile point of CP CDF), the cost ($Cost_{0.1}^{99}$ based on Eq. 13), and GAR degradation ($\Delta \mathcal{R}_{0.1}^{99}(10)$ based on Eq. 12). As the results show, using higher Vth results in lower GAR degradation but a higher cost and higher initial delay. Moreover, it is observed that when DVth assignment is used selectively (only on the NAND 5 gate), lifetime reliability degradation is reduced and imposed cost decreases, thus providing a reliability-cost tradeoff. Hence, if appropriate gates are selected in applying the DVth assignment technique, the lifetime reliability of the circuit can be improved by controlling the imposed cost. On the other hand, for a combinational circuit with n gates, the number of cases for applying the DVth assignment technique is 2^n; i.e., each gate can be assigned with either low or high Vth. So applying the DVth assignment technique to combinational circuits is not a straightforward task. In the next section, we introduce an optimization flow and then present three different algorithms to be used for the improvement of the lifetime reliability of combinational circuits using the DVth assignment technique.

5   Lifetime Reliability Optimization Flow

The overall view of the presented lifetime reliability optimization flow is shown in Fig. 4. Based on HSPICE simulations, BTI model, and PV effect information, the gate delay model is constructed and then incorporated into the aging-aware SSTA, taking into account the duty cycle calculations. Based on SSTA results, a PV- and BTI-aware criticality metric is used to identify the statistically critical gate set as the candidates for applying the DVth assignment technique. Then the DVth technique is applied to (one or more) critical gate(s) in the circuit based on the optimization algorithm's policy, i.e., the greedy-based method (GeRO), the simulated annealing (SA)-based method (SARO), and the sensitivity-based (TIRO) method. Then the critical gate set is updated, and this flow continues until the critical gate set is empty. The algorithm always terminates as there are only two cases for each gate; i.e., it can be with either low Vth or high Vth.

In the following, we present first the PV- and BTI-aware criticality metric and then the three different approaches for the improvement of lifetime reliability.

5.1   Process Variation- and BTI-Aware Criticality Metric

In order to identify the candidate gates for applying the DVth assignment technique, a PV- and BTI-aware criticality metric is presented. The criticality of gate i is defined as:

$$Criricality(i) = \frac{\partial D_i}{\partial Vth_i} \times Slack(i) \tag{14}$$

where $\frac{\partial D_i}{\partial Vth_i}$ is the Vth variation-induced gate delay deviation and $slack(i)$ is the gate delay slack, which is computed as:

$$Slack_{(Gate\ i)} = RAT_{(Gate\ i)} - AT_{(Gate\ i)} \tag{15}$$

where RAT and AT, respectively, show the required arrival time and arrival time for gate i. Whatever Vth variation-induced gate delay variation is more significant for a gate i, its delay degrades more than other gates. So picking this gate out is more appropriate for dual Vth. On the other hand, whatever gate delay slack is more significant for gate i, assigning high Vth results in fewer negative impacts on circuit initial delay. So this gate is more suitable for dual Vth in terms of timing yield loss. Thus, whatever *criricality*(i) is higher, the gate is more affected by PV and BTI applying the DVth assignment technique reduces the variation effect while less performance overhead.

5.2   Optimization Algorithm-Based DVth Assignment Policy

In order to improve the lifetime reliability of digital circuits, we take advantage of different DVth assignment policies inspired by three optimization approaches. The main difference in these optimization approaches is the provided tradeoff between the reliability improvements and the optimization speed. The details of these approaches are presented in the following.

5.2.1   Optimization Approach #1: Greedy-Based Method (GeRO)

Algorithm 1 shows the pseudocode of the greedy-based lifetime reliability optimization method (GeRO). After computing the criticality of each gate, the gates are ranked in descending order based on the computed criticality described in Eq. 14. Then the most critical gate is picked out for assigning high Vth. Next, the circuit timing is computed using an incremental SSTA; i.e., only the delay information of the gates in the fan-in cone of the modified gate is recalculated, resulting in less computation time. In order to evaluate the impacts of the DVth assignment technique, the GAR metric described in Sect. 1.4 is used. If the GAR degradation value of the new circuit is less than the old one and the cost value computed based on Eq. 13 is less than the user-specific constraint, a high Vth assignment to the current gate is considered an acceptable *move* in the algorithm, and the criticality list will be updated; otherwise, the circuit is rollbacked (i.e., the selected gate Vth is changed back to low Vth), and the gate is removed from the critical list. The process is repeated until the critical list becomes empty.

Algorithm 1: GeRO Method
Inputs: netlist, ordered candidate list (C_e), overhead constraint
Output: optimized circuit

1. For each gate i in the C_e list

    1.1. Assign high Vth to gate i
    1.2. Recompute the timing of the circuit incrementally
    1.3. If cost < user constraint and GAR degradation of the new circuit < GAR degradation of the previous circuit

        1.3.1. Accept the new solution
        1.3.2. Terminate the optimization policy

    1.4. Else

        1.4.1. Reject the new solution and mark it as an impossible candidate

2. End

5.2.2 Optimization Approach #2: Simulated-Annealing-Based Method (SARO)

The pseudocode of the presented simulated-annealing-based reliability optimization method (SARO) algorithm is presented in Algorithm 2. The algorithm starts by assigning low Vth to all gates in the circuit as the initial solution (x_c). Then the criticality of all gates is computed using Eq. 14, and the most critical gate is picked out for high Vth assignment. Afterward, the circuit timing is computed incrementally for the modified circuit, called the new solution (x_n). If the new solution is better in terms of GAR degradation, the new solution be accepted and the current solution will be replaced by the new one. Otherwise, if the new solution is worse, it may be still accepted if a randomly generated number between 0 and 1 is higher than the probabilistic value p, computed as:

$$p = e^{\left(\frac{GAR\ deg(x_c) - GAR\ deg(x_n)}{temp}\right)} \qquad (16)$$

where $GAR\ deg\ (x_c)$ and $GAR\ deg\ (x_n)$ are, respectively, the GAR degradation of the solution (c) and the new solution (n), which is computed using Eq. 12, and $temp$ is the temperature used to determine the acceptance probability of the worse solution. To avoid premature convergence, the rate of $temp$ reduction tends to zero [22] as:

$$T_{i+1} = \alpha * T_i \qquad (17)$$

where i is the number of iterations and α is the cooling rate ($0 < \alpha < 1$).

Accepting worse new solutions occurs more at the beginning of the SARO algorithm (due to high value of $temp$) for avoiding local optimum solutions [22]. However, the value of $temp$ is reduced in the next iterations (based on Eq. 17), and thus, only the improved ones are allowed, resulting in faster algorithm convergence. Further information on SARO can be found in the survey work in Ref. [22]. The SARO algorithm terminates until there is no critical gate in the candidate set. Since the critical list size is limited (total gates in circuits at maximum), it guarantees that the SARO optimization algorithm will finally converge.

Algorithm 2: Simulated-Annealing-Based Lifetime Reliability Optimization Method

Inputs: netlist, ordered candidate list (C_e), overhead constraint
Output: optimized circuit

1. Set initial solution x_c and temperature
2. For each gate i in C_e list

    2.1. Assign high Vth to gate i
    2.2. Recompute the timing of the circuit incrementally (new solution x_n)
    2.3. If probability in Eq. 6 > random (0, 1)

        2.3.1. Accept the new solution
        2.3.2. Terminate the optimization policy

    2.4. Else

        2.4.1. Reject the new solution and mark it as an impossible candidate

3. End

5.2.3 Optimization Approach #3: Sensitivity-Based Method (TIRO)

In this section, we present a sensitivity-based lifetime reliability optimization algorithm using the DVth assignment technique. In the presented sensitivity-based optimization algorithm, the threshold voltage of gates is incrementally increased one by one to determine the threshold voltage assignment that provides the best reliability value. For each gate, it is important to consider the relative reliability improvement of a given gate to the imposed cost of the optimization technique. So we present a sensitivity metric to measure and analyze the impact of the dual Vth of the gates on the reduction of the variation effect of the whole circuit. At first, the critical set is obtained based on Eq. 14, and then for each gate in the critical list, the timing information is updated using the aging-aware statistical timing analysis. Then the best gate, one that provides the most benefit (best GAR degradation improvement) to the imposed cost (the lowest delay overhead), is selected based on the sensitivity metric, which is computed as:

$$Sensitivity_{(gate\ i)} = \frac{GAR\ deg.imp.(i)}{cost(i)} \qquad (18)$$

where *cost(i)* is computed based on Eq. 13 when high Vth is assigned to gate *I*, and *GAR deg . imp.* (*i*) shows the GAR degradation improvement and is computed as:

$$GAR\ deg.imp. = \frac{\Delta \mathcal{R}_g^p(t)_{Before} - \Delta \mathcal{R}_g^p(t)_{After}}{\Delta \mathcal{R}_g^p(t)_{Before}} \qquad (19)$$

where $\Delta \mathcal{R}_g^p(t)_{Before}$ and $\Delta \mathcal{R}_g^p(t)_{After}$ show the reliability degradation of the circuit before and after applying the DVth technique (assigning high Vth to a chosen gate), respectively.

After assigning high Vth to the gate with the highest sensitivity, the critical list is updated. The optimization loop repeats until the critical list becomes empty. The pseudocode sensitivity-based lifetime reliability optimization algorithm is presented in Algorithm 3.

Algorithm 3: TIRO Method
Inputs: netlist, ordered candidate list (C_e), overhead constraint
Output: optimized circuit

1. For each gate *i* in C_e list

   1.1. Compute the sensitivity of gate *i* (Eq. 18)

2. Sort the sensitivity list
3. Assign high Vth to the gate with the most sensitivity and accept this solution
4. End

---

6   Experimental Results

In this section, we conduct a set of experiments to investigate the different aspects of the presented approaches for the improvement of lifetime reliability. Firstly, we study the error of the presented aging-aware delay model, and then the reliability of ISCAS'85 and ISCAS'89 benchmark suits are analyzed in the presence of aging and PV. Then the lifetime reliability optimization results for different optimization algorithms are investigated.

The presented approaches are implemented in C++ and are run on a windows machine with a core i7 quad-core Intel processor (4.6 GHz) and 32-GB random access memory (RAM). The presented approach was applied to ISCAS'85 and ISCAS'89 benchmark circuits. In this chapter, the primitive gates (i.e., INV, BUFF, 2- to 4-input NAND, and 2- to 4-input NOR) are used in the netlist synthesis process. It is notable that ISCAS'89 benchmark circuits are sequential circuits; i.e., they have a combinational part and storage element (such as flip flops (FFs)). So these circuits are converted into combinational ones by removing FFs and adding the FFs' inputs/outputs to the primary outputs (POs)/primary inputs (PIs) of the circuit.

Table 2 ISCAS'85 and ISCAS'89 benchmark circuit information

Bench.	# Gates	# Flip flops	(#PI, #PO)
C880	383	–	(60,26)
C2670	1193	–	(233,140)
C5315	2307	–	(178,123)
C7552	3512	–	(207,108)
S420	196	16 D-FF	(19,2)
S820	289	5 D-FF	(18,19)
S1488	653	6 D-FF	(8,19)
S5378	2779	179 D-FF	(35,49)
S15850	9772	534 D-FF	(77,150)

Table 3 Error of the PV- and BTI-aware gate delay model

	Inv	NAND			NOR		
Error (%)		2	3	4	2	3	4
σ	1.02	1.82	2.07	3.11	1.46	2.48	3.47
μ	1.36	2.39	2.61	3.42	3.22	3.74	3.98

Table 2 provides information about the benchmark circuits used in the following experiments. The first and second columns show the circuit name and the number of primitive gates in benchmarks. The third column shows the number of FFs in the ISCAS'89 benchmark. The dashed line in some rows indicate that ISCAS'85 benchmark circuits have no FF. The last column presents the number of PIs and POs.

The value of the fitting parameters A, B, γ, and β in the statistical aging model are computed using HSPICE simulations and MOSFET model reliability analysis (MOSRA) [23]. HSPICE simulations are accomplished through the PTM 22 nm technology model [24], supply voltage 1.1V, and 354K temperature.

To model the spatial correlation, a three-level quad-three partition is employed [25, 26]. All the gates are distributed randomly in a 4×4 grid at the bottom level. Then the random variables of the gate delays are computed depending on the hierarchy of the gate. Based on Ref. [25], the random variables at the same level have the same probability distribution. In this work, we use a different total PV effect on Vth, i.e., 5%, 7%, and 10%. For the 10% PV, it is split into two parts, i.e., the first part for considering the systematic variance of 6% and the other one for the random variance of 8%.

6.1 PV- and BTI-Aware Delay Degradation Model Verification

In order to verify the accuracy of the presented PV- and BTI-aware gate-level delay degradation model, we compare the gate delay values obtained from Eq. 6 with the results of the Monte-Carlo-based HSPICE (MCH) simulation, which was performed under the same variation distribution and working conditions. Table 3 presents the relative error values between the presented delay degradation model and MCH simulation for the primitive gates. As the results show, the maximum error on the μ and σ values of gate delay between the presented model and MCH simulation is less than 4%. It is notable that the error originated in the parameter fitting process. Despite this error, the PV- and BTI-aware delay model has adequate accuracy to be used in a reliability improvement framework under PV and BTI effects.

6.2 Lifetime Reliability Analysis

In order to investigate the severity of the lifetime reliability challenge in the nanoscale digital circuit, the presented statistical PV- and BTI-degradation model is incorporated into an SSTA to calculate μ and σ of circuit maximum arrival times. The analysis is performed considering various variation ratios and different operational times. The duty cycles of the signals are computed considering their signal probabilities. The SP values of all primary inputs (PIs) are set to 0.5, and for internal nodes, SP is calculated using the approach presented in Ref. [27]. All the gates in the circuit are initially assigned with a low Vth.

Figure 5 shows the CP delay values (computed as $\mu + 3\sigma$ from the CP delay distribution) after 3, 6, and 9 years of operation time normalized to design CP delay. As expected, the delay values are increased (about 2× in most benchmark circuits after 9 years) due to BTI effects by increasing circuit operation time. So BTI poses a severe challenge for satisfying the timing constraints of the digital circuits in the operational lifetime.

The impacts of PV on the lifetime reliability of digital circuits are also investigated. Figure 6 shows GAR for variation ratios of 5%, 7%, and 10%. As shown in the figure, by increasing the PV amount (expected in future technology nodes), GAR is also decreased, indicating that lifetime reliability improvement should be addressed taking into account the impacts of PV on digital circuits.

Fig. 5 Normalized CP delay during lifetime

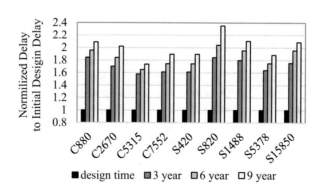

■design time   ■3 year   ■6 year   □9 year

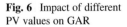 **Fig. 6** Impact of different PV values on GAR

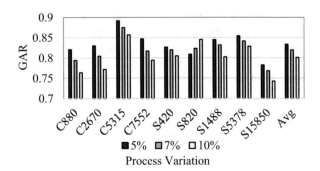

Table 4 Initial GAR degradation of the benchmark circuits (before applying any optimization method)

Bench.	Variation ratio								
	5%			7%			10%		
	Lifetime (y)			Lifetime (y)			Lifetime (y)		
	3	6	9	3	6	9	3	6	9
C880	13.82	17.93	23.68	15.49	20.57	26.84	18.82	23.59	29.30
C2670	12.01	17.02	23.08	14.51	19.55	25.13	16.41	22.80	29.73
C5315	7.58	10.79	13.84	9.29	12.46	15.35	11.91	14.27	17.81
C7552	9.62	15.26	20.83	12.18	18.29	23.21	14.21	20.55	24.86
S420	10.32	17.35	26.93	10.93	18.02	26.78	12.36	19.48	28.39
S820	9.85	14.90	19.22	11.02	17.65	22.97	13.16	19.08	25.42
S1488	10.84	15.47	21.06	11.90	16.76	22.56	14.46	19.74	25.32
S5378	6.95	9.90	12.64	10.33	15.81	19.78	12.01	17.12	21.77
S15850	15.41	21.72	28.52	15.75	23.14	29.19	18.34	25.72	31.96

6.3   Lifetime Reliability Optimization

In order to show the efficacy of the presented approaches for the improvement of lifetime reliability (GeRO, SARO, and TIRO methods), the reliability of ISCAS'85 and ISCAS'89 benchmark circuits are evaluated under different variation ratios, operational lifetime, and guardband values.

At first, GAR degradation of the circuits under BTI effects (considering 3-, 6-, and 9-year operation time) and PV effects (considering 5%, 7%, and 10% variation) before applying any optimization technique is presented in Table 4. As the results show, GAR degradation is increased during the circuits' lifetime; for example, the GAR degradation of the C880 benchmark circuit is, respectively, 13.82%, 17.93%, and 23.68%, considering operation times of 3-, 6-, and 9 years under a 5% PV. Also, increasing the PV effects leads to an increase in GAR degradation; for example, the GAR degradation of the C880 benchmark using 5%, 7%, and 10% variation ratios are, respectively, 13.82%, 15.49%, and 18.82% for a 3-year lifetime.

Table 5 GAR degradation optimization results obtained by the GeRO algorithm

	Variation ratio								
	5%			7%			10%		
	Lifetime (y)			Lifetime (y)			Lifetime (y)		
Bench.	3	6	9	3	6	9	3	6	9
C880	8.23	12.69	14.66	9.49	14.03	16.63	12.57	17.00	22.66
C2670	7.91	11.85	16.49	9.64	14.25	18.22	11.34	16.04	22.51
C5315	4.82	6.29	8.15	5.91	7.63	9.47	7.07	8.98	11.32
C7552	6.15	9.03	10.99	7.77	11.18	13.02	9.38	12.88	14.50
S420	4.19	7.77	11.56	5.57	10.02	14.96	7.60	12.11	18.97
S820	7.84	10.63	15.48	9.56	12.74	17.12	9.85	14.90	19.20
S1488	7.43	10.20	13.61	8.88	11.38	15.36	10.47	14.35	18.36
S5378	4.25	6.38	9.04	5.17	7.61	10.87	6.95	9.90	12.64
S15850	7.47	12.99	16.90	8.44	14.21	18.28	10.80	16.35	21.90

Table 6 GAR degradation optimization results obtained by the SARO algorithm

	Variation ratio								
	5%			7%			10%		
	Lifetime (y)			Lifetime (y)			Lifetime (y)		
Bench.	3	6	9	3	6	9	3	6	9
C880	7.14	10.90	13.86	8.34	12.11	15.76	11.64	15.79	19.69
C2670	5.77	9.64	13.97	7.07	11.38	16.64	8.61	13.72	19.68
C5315	6.00	7.65	10.09	7.12	8.83	11.38	8.79	10.06	13.02
C7552	5.67	7.41	9.25	6.49	9.71	11.97	8.55	11.85	13.33
S420	4.19	7.77	11.56	5.57	10.02	14.96	7.60	12.11	18.97
S820	4.68	7.28	10.93	5.78	8.67	12.18	7.31	10.72	14.54
S1488	5.08	9.22	12.47	6.24	10.83	14.22	7.42	12.30	15.54
S5378	4.60	7.89	10.68	5.67	8.96	11.27	7.08	10.10	13.14
S15850	5.79	9.13	13.74	6.65	10.39	14.46	8.06	12.31	15.53

Tables 5, 6 and 7 respectively show the GAR degradation values obtained from GeRO, SARO, and TIRO reliability improvement algorithms under BTI effects (considering 3, 6, and 9 years of operational times) and PV effects (considering 5%, 7%, and 10% variations) with 10% guardband. As shown in the tables, the presented DVth technique incorporated into the different optimization approaches leads to the improvement of the lifetime reliability of digital circuits as operational lifetime and PV effects increase. For example, in the GeRO approach, the GAR degradation of the C880 benchmark for a 6-year lifetime and 5% PV is improved from 17.93% to 12.69% after optimization. Also, GAR degradations using SARO and TIRO optimization methods are reduced to 10.90% and 7.77%, respectively. In all cases, TIRO has the best improvement as compared to the GeRO and SARO methods. GeRO and SARO assign high Vth to the most critical gate, which is selected based on the criticality metric, taking into account delay slack as well as Vth sensitivity. So a gate with a big slack and medium Vth sensitivity may become

Table 7 GAR degradation optimization results obtained by the TIRO algorithm

Bench.	Variation ratio								
	5%			7%			10%		
	Lifetime (y)			Lifetime (y)			Lifetime (y)		
	3	6	9	3	6	9	3	6	9
C880	4.89	7.77	10.14	6.33	9.38	11.49	8.21	11.95	14.30
C2670	4.50	6.49	10.71	5.98	8.12	12.73	7.30	10.57	16.96
C5315	3.12	5.49	6.51	4.08	6.32	7.76	5.22	7.44	9.96
C7552	4.17	6.78	8.45	5.30	8.30	10.20	6.57	9.59	11.93
S420	4.19	7.77	11.56	5.57	10.02	14.96	7.60	12.11	18.97
S820	3.51	6.46	9.68	4.79	7.75	11.19	6.19	8.51	13.52
S1488	4.06	6.04	8.37	5.28	7.57	10.24	6.32	9.08	12.10
S5378	2.12	4.20	7.27	3.09	5.13	8.34	4.96	7.10	10.12
S15850	3.08	4.90	6.39	4.04	6.02	7.65	5.08	7.10	8.93

the most critical gate in the GeRO and SARO methods and, thus, be selected as the most critical gate in all cases. Hence, assigning high Vth to the most critical gate in GeRO and SARO methods does not necessarily result in achieving the most improvement in terms of reliability. On the other hand, the TIRO approach computes the sensitivity metric, which is the relative reliability improvements per delay overhead achieved by assigning high Vth to each critical gate (i.e., in each optimization iteration of TIRO, all critical gates are analyzed and the gate with the most reliability improvement and the least delay overhead is assigned with high Vth). So TIRO directly considers reliability improvement and finds the gates for applying high Vth more intelligently and more efficiently.

6.4 Algorithm Computation Complexity and Runtime

An important aspect of an optimization algorithm in computer-aided designs is its computation complexity and runtime. The computation complexity of GeRO and SARO is $O(n)$, where n is the total number of gates in the combinational circuit. Because GeRO and SARO accept or reject the current critical gate for applying the DVth technique, this gate will not appear in the next critical gate list. When the list becomes empty, the optimization is terminated, and thus, the optimization will be terminated at $O(n)$. On the other hand, TIRO firstly computes the reliability improvement per cost, which is achieved by assigning high Vth to each gate in the critical list. So it is required to analyze reliability and technique cost for all the gates in critical gates; i.e., the circuit is analyzed for all the critical gates in each iteration. As a result, TIRO has $O(n^2)$ time complexity. However, the results show that the number of critical gates is very much less than the total number of gates in the circuits.

Table 8 Runtime comparison of the presented optimization algorithms

Bench.	Gates (PI, PO)	Runtime (s)		
		GeRO	SARO	TIRO
C880	(60,26)	10.4	10.6	100.9
C2670	(233,140)	351.2	342.3	1274.2
C5315	(178,123)	864	845	2709
C7552	(207,108)	1280	1211	3475
S420	(19,2)	1	1.2	9.2
S1488	(8,19)	12.4	9	63
S5378	(35,49)	990	964	2177
S15850	(150,77)	11990	12543	46532
Average	–	1626	1665	5834

Table 8 shows the runtime of the presented optimization algorithms for different benchmark circuits. The first two columns show the name and information (number of gates, PIs, and POs) of the benchmark circuits, and the other three columns respectively represent the runtime of the GeRO, SARO, and TIRO methods. It is observed that the runtime of the GeRO and SARO methods are approximately equal (on average, GRO takes 1626 s and SARO takes 1665 s to finish), while TIRO has a 5834-s runtime. As can be seen, GeRO and SARO have similar runtimes, while TIRO takes longer to optimize the circuit.

7 Conclusion

There have been major challenges to the lifetime reliability of nanoscale digital circuits due to PV and BTI effects. This chapter presents a statistical circuit optimization framework to optimize circuit lifetime reliability under the joint effect of PV and BTI. Based on a novel metric for a lifetime reliability evaluation, circuit reliability is improved by assigning high Vth to candidate gates, which results in reducing BTI effects while increasing initial delay at the design time, i.e., the timing yield overhead. The reliability improvement procedure is developed using three different optimization algorithms, i.e., greedy-based (GeRO), SA-based (SARO), and TILOS-like sensitivity-based (TIRO) algorithms. The experimental results show averagely 6.38%, 8.16%, and 9.93% reliability improvement for the GeRO, SARO, and TIRO optimization algorithms, respectively, while imposing 6.50%, 7.03%, and 6.90% timing yield overhead, respectively. The obtained results show that TILOS achieves the best reliability improvement with higher computation time.

There are many avenues as the future work of this chapter. First, combining techniques such as gate resizing with dual Vth can be beneficial in achieving more reliability improvement as gate resizing decreases the DVth timing yield overhead. Moreover, other evolutionary optimization algorithms (such as a genetic algorithm) can be applied to gain higher circuit reliability improvement for combinational circuits.

References

1. Bowman KA, Duvall SG, Meindl JD. Impact of die-to-die and within-die parameter fluctuations on the maximum clock frequency distribution for Gigascale integration. IEEE J. Solid-State Circuits. 2002;37(2):183–90.
2. Jafari A, Raji M, Ghavami B. Impacts of process variations and aging on lifetime reliability of flip-flops: a comparative analysis. IEEE Trans Device Mater Reliab. 19(3):551–62. https://doi.org/10.1109/TDMR.2019.2933998.
3. Agarwal A, Paul BC, Mahmoodi H, Datta A, Roy K. A process-tolerant cache architecture for improved yield in nanoscale technologies. IEEE Trans Very Large Scale Integr Syst. 2005;13(1):27–37.
4. Grossi M, Omaña M. Impact of Bias Temperature Instability (BTI) aging phenomenon on clock deskew buffers. J Electron Test Theory Appl. 2019;35:261.
5. Jafari A, Raji M, Ghavami B. Timing reliability improvement of master-slave flip-flops in the presence of aging effects. IEEE Trans Circuits Syst I Regul Pap. https://doi.org/10.1109/TCSI.2020.3024601
6. Jin S, Han Y, Li H, Li X. Statistical lifetime reliability optimization considering joint effect of process variation and aging. Integr VLSI J. 2011;44(3):185–91.
7. Khan S, Hamdioui S, Kukner H, Raghavan P, Catthoor F. Incorporating parameter variations in BTI impact on nano-scale logical gates analysis. In: Proceeding of the IEEE international symposium on Defect and Fault Tolerance in VLSI and Nanotechnology Systems; 2012. p. 158–63.
8. Lu Y, Shang L, Zhou H, Zhu H, Yang F, Zeng X. Statistical reliability analysis under process variation and aging effects. In: 2009 46th ACM/IEEE design automation conference; 2009. p. 514–9.
9. Firouzi F, Kiamehr S, Tahoori MB. Statistical analysis of BTI in the presence of process-induced voltage and temperature variations. In: 2013 18th Asia and South Pacific Design Automation Conference (ASP-DAC); 2013. p. 594–600.
10. Ebrahimipour SM, Ghavami B, Raji M. A statistical gate sizing method for timing yield and lifetime reliability optimization of integrated circuits. IEEE Trans Emerg Top Comput. 2021;9(2):759–73. https://doi.org/10.1109/TETC.2020.2987946.
11. Ghavami B, Raji M, Rasaizadi R, Mashinchi M. Process variation-aware gate sizing with fuzzy geometric programming. Comput Electr Eng. 2019;78:259–70.
12. Raji M, Sabet MA, Ghavami B. Soft error reliability improvement of digital circuits by exploiting a fast gate sizing scheme. IEEE Access. 2019;7:66485–95. https://doi.org/10.1109/ACCESS.2019.2902505.
13. Gomez A, Champac V. An efficient metric-guided gate sizing methodology for guardband reduction under process variations and aging effects. J Electron Test. 2019;35(1):87–100.
14. van Santen VM, Amrouch H, Martin-Martinez J, Nafria M, Henkel J. Designing guardbands for instantaneous aging effects. In: Design Automation Conference (DAC), 2016 53nd ACM/EDAC/IEEE; 2016. p. 1–6.
15. Wang Y, Luo H, He K, Luo R, Yang H, Xie Y. NBTI-aware dual Vth assignment for leakage reduction and lifetime assurance. Chinese J Electron. 2009;18(2):225–30.
16. Jafari A, Raji M, Ghavami B. BTI-aware timing reliability improvement of pulsed flip-flops in nano-scale CMOS technology. IEEE Trans Device Mater Reliab. 2021;21(3):379–88. https://doi.org/10.1109/TDMR.2021.3102521.
17. Yang S, Wang W, Hagan M, Zhang W, et al. NBTI-aware circuit node criticality computation. ACM J Emerg Technol Comput Syst. 2013;9:Article No.: 23. dl.acm.org
18. Raji M, Mahmoudi R, Ghavami B, Keshavarzi S. Lifetime reliability improvement of nano-scale digital circuits using dual threshold voltage assignment. IEEE Access. 2021;9:114120–34. https://doi.org/10.1109/ACCESS.2021.3103200.
19. Fishburn JP, Dunlop AE. Tilos: a posynomial programming approach to transistor sizing. In: Proceeding of the IEEE/ACM ICCAD; 1985. p. 326–8.

20. Duan S, Halak B, Zwolinski M. An ageing-aware digital synthesis approach. In: 2017 14th international conference on Synthesis, Modeling, Analysis and Simulation Methods and Applications to Circuit Design (SMACD); 2017. p. 1–4.
21. Rao VG, Mahmoodi H. Analysis of reliability of flip-flops under transistor aging effects in nano-scale CMOS technology. In: 2011 IEEE 29th international conference on Computer Design (ICCD); 2011. p. 439–40.
22. Suman B, Kumar P. A survey of simulated annealing as a tool for single and multiobjective optimization. J Oper Res Soc. 2006;57(10):1143–60.
23. Maricau E, Gielen G. Analog IC reliability in nanometer CMOS. New York: Springer; 2013.
24. N. I. and M. (NIMO) Group. Predictive Technology Model (PTM), 2005. [Online]. Available: http://ptm.asu.edu/. Accessed 16 Jan 2020.
25. Raji M, Ghavami B, Zarandi HR, Pedram H. Process variation aware performance analysis of asynchronous circuits considering spatial correlation. In: Proceedings of the 19th international conference on integrated circuit and system design: power and timing modeling, optimization and simulation (PATMOS'09). Berlin/Heidelberg: Springer. p. 5–15.
26. Agarwal A, Blaauw D, Zolotov V. Statistical timing analysis for intra-die process variations with spatial correlations. In: Proceedings of the 2003 IEEE/ACM international conference on Computer-aided design; 2003. p. 900.
27. Ercolani S, Favalli M, Damiani M, Olivo P, Ricco B. Estimate of signal probability in combinational logic networks. In: [1989] Proceedings of the 1st European test conference; 1989. p. 132–8.

Index

Printed in the United States
by Baker & Taylor Publisher Services